Optical Data Storage

Philips Research

VOLUME 4

Editor-in-Chief
Dr. Frank Toolenaar
Philips Research Laboratories, Eindhoven, The Netherlands

SCOPE TO THE *'PHILIPS RESEARCH BOOK SERIES'*

As one of the largest private sector research establishments in the world, Philips Research is shaping the future with technology inventions that meet peoples' needs and desires in the digital age. While the ultimate user benefits of these inventions end up on the high-street shelves, the often pioneering scientific and technological basis usually remains less visible.

This 'Philips Research Book Series' has been set up as a way for Philips researchers to contribute to the scientific community by publishing their comprehensive results and theories in book form.

Dr. Rick Harwig

Optical Data Storage
Phase-Change Media and Recording

By

Erwin R. Meinders
Philips Optical Media & Technology, Eindhoven, The Netherlands

Andrei V. Mijiritskii
Philips Lighting, Business Unit Automotive Lighting, Eindhoven,
The Netherlands

Liesbeth van Pieterson
Philips Research, Eindhoven, The Netherlands

and

Matthias Wuttig
I. Physikalisches Institut, RWTH Aachen, University, Aachen, Germany

A C.I.P. Catalogue record for this book is available from the Library of Congress.

ISBN-10 1-4020-4216-7 (HB)
ISBN-13 978-1-4020-4216-4 (HB)
ISBN-10 1-4020-4217-5 (e-book)
ISBN-13 978-1-4020-4217-1 (e-book)

Published by Springer,
P.O. Box 17, 3300 AA Dordrecht, The Netherlands.

www.springer.com

Printed on acid-free paper

Contents

1. Introduction

1.1.   A brief overview of optical storage systems

Today's optical storage system stems from a small-scale product developed by Philips and commercially launched in 1978. This system was the result of the Videodisc project that was running at the Philips Research labs in Eindhoven, The Netherlands, through the 1970s. [1] It was pioneering laser-based optical storage and was based on an analogue videodisc. The product never broke the boundaries of its market niche and at its decline the number of contents titles was quite limited. However, with its optical pick-up head, servo electronics, disc mastering principles, and fine mechanics it formed a basis for the optical storage technology employed nowadays. Unlike its predecessor, the next generation system was to revolutionize the world of data storage. The fruits of a close collaboration between Philips and Sony were officially made public in 1979 in the form of a worldwide standard. The first product became commercially available in 1983 under the name of Compact Disc (CD). This was a shiny 12 cm disc carrying about 74 minutes of music in a digital format. [2] Fostered by the fast growth of computer industry a CD for computer applications – Compact Disc Read-Only Memory (CD-ROM) was introduced on the market in 1985. The disc could hold up to 650 megabyte (MB) of data and at 1x disc speed the data transfer rate was 4.3 Megabit per second (Mbps). The CD-ROM makes use of the same physical format as CD-audio but has additional error detection and correction encoding. To meet the ongoing developments in multimedia applications a number of derivatives from the original CD-ROM format have been added to the CD family. Most prominent of them were CD-interactive (CD-I) and video-CD. The CD-I format was defined to enable computer-based digital storage of data, audio, graphics and video. Video-CD is used to store 74 min of combined full-motion video and audio employing MPEG-1 video data compression techniques. Following the success of read-only discs, recordable (CD-R) and rewritable (CD-RW) media completed the family of first-generation optical disc storage in 1984 and 1995, respectively. [3] An important feature of the CD family is the high degree of interchangeability between its different family members. This was one of the main factors promoting the success of the CD optical storage system.

The tremendous technological developments of the late 80s and early 90s have created a great demand and a suitable technological basis for higher data capacities and data rates. In 1996, the second-generation optical storage system – Digital Versatile Disc (DVD) was launched. The disc accommodates 4.7 GB of data on one data layer and its DVD-video format delivers about 2.5 hours of standard-definition (SD) digital video. DVD makes use of the Universal Disc Format (UDF) to enable multimedia applications in both consumer electronics appliances and computer peripherals. It employs MPEG-2 for video compression. At 1x disc speed the system provides a data transfer rate of 11 Mbps. Besides single-layer discs, dual-layer and double-sided dual-layer configurations have been developed, with 8.5 GB and 17 GB of data capacities, respectively. Within a few years, recordable and rewritable DVD media have appeared on the market. Three mutually incompatible formats – DVD-RAM, DVD-R/RW, and DVD+R/RW, have been standardized by different industry alliances. This incompatibility has led to a so-called format war, which left both the industry players and the consumer on the loosing side. [2], [4]

In June 2002, standardization of a third-generation optical storage system was finalized. The system is called Blu-ray Disc (BD) and was proposed by the Blu-ray Disc Founders, an industrial consortium of 9 leading companies (9C consortium), comprising, among others LG, Samsung, MEI, Sony, Philips, Thomson, Hitachi, TDK, etc. [4], [5] The BD system evolved from the DVR (Digital Video Recording) project running at Philips Research labs and Sony since 1996, this time pioneering blue laser recording. [6] Many of the physical parameters proposed in the DVR system were also adapted to the BD system. The BD system features 25 GB single-layer and 50 GB dual-layer 12 cm discs and a data transfer rate of 36 Mbps. Two other data capacities are also described in the format, 23.3 GB and the reserved 27 GB. In contrast to the preceding generations, it was the rewritable disc format (BD-RE) that was described in the first version of the standard. High-definition (HD) video recording is anticipated to be the main driving force from the application side. In 2003, Sony launched the first commercial BD video recorder in Japan. Only BD-RE discs of 23.3 GB data capacity (according to version 1.0 of the BD-RE book) can be used on this recorder. In the mean time, almost all 9C companies started their own drive development activities. A major breakthrough in the proliferation of BD is the successful introduction of the triple-writer optical pick-up unit (OPU) by Philips, ensuring backwards compatibility up to the first generation optical discs. This OPU can actually read and **write** CD, DVD and BD type of discs. Next to Blu-ray Disc, another third-generation system has been proposed. This system is currently being standardized under the name of HD-DVD (high-definition DVD) and has some major physical differences from BD. A main difference with the BD system is the lower data capacity of 15 GB; a dual-layer version makes 30 GB storage capacity. The future of both systems is unclear as yet. One of the serious issues the optical storage industry has to deal with is copy protection (CP) and digital rights management (DRM) of the data stored on the discs. There is a high probability that success of the upcoming generations of optical storage systems will be determined not only by their storage and retrieval performance but to a large extent by the availability and versatility of CP/DRM solutions.

So, by the time this book is being written two generations of optical storage systems and a plurality of often competing formats have been successfully commercialized. The 'war' on the third generation 'blue' systems has just started. It seems that the first recordable high-density systems will be utilizing the BD format with a disc capacity of up to 50 GB. Foreseen applications are a high-density video recorder and a PC drive. The high-density format allows also for a smaller-form factor drive, such as a Camcorder. The availability of BD-ROM media is of strategic importance for the proliferation of the BD format. But the willingness of leading film studios and content distributors to publish high-definition content in BD-ROM format depends very much on their confidence in the copy protection system of BD-ROM.

While the market introduction of the third generation optical recording system just started, options for a fourth generation system are already under development in the research labs of several companies. In accordance to the evolutionary increase in data capacity, the near-field system utilizes an increased numerical aperture objective lens to allow for a 100GB single layer data capacity. [7] Also advanced signal processing widens the system margins and enables a single-layer storage capacity of up to 50 GB. [8] Two-dimensional optical storage is a possibility to increase data transfer rates but this system requires a multi-pot readout system. [9] Data capacities of Terabytes are envisioned if the third dimension is explored, so-called volumetric data storage. Recent improvements in recording materials have renewed interest in holographic data storage. [10] Besides the tremendous data capacity, page-based storage involves also a relative high data-transfer rate. Other examples of volumetric data storage are electrochromic media, in which the individual data layers are independently addressable. [11] The University of Arizona explores currently an evolutionary optical storage system based on DNA carriers. [12] Although this system is far away from commercialization, it is based on a very interesting and novel concept.

Nowadays an attractive property portfolio, which includes removeability, robustness, interchangeability, low price, characterizes optical storage media and 'cool' look. But what is the physical difference between the different storage generations and how does it work all together? In what follows the principles of optical data storage will be explained on a basic level.

1.2.   The basics of optical storage

1.2.1.   Optical drive layout

An optical storage system consists of an optical drive and corresponding optical media. The main elements of an optical drive are a semiconductor laser, a set of optical elements to shape and focus the laser beam, a disc driving part, and a signal detection system. In Figure 1, a simplified layout of an optical drive is shown. A light beam generated by the laser propagates through the optical elements of the drive and is focused into a diffraction-limited spot on the disc. Being reflected by the disc, which carries user and service information, the beam is projected onto a set of photo-detectors. The detected signals are subsequently processed by electronics of the drive (not shown in the figure). Among the most important parameters that characterize an optical storage drive are the wavelength (λ) of the laser and the numerical aperture of the objective lens. The numerical aperture is defined as $NA = \sin\alpha$, where α is the angle between the optical axis and the marginal ray of the converging beam in air. As will be shown below, these parameters determine the storage density of the system.

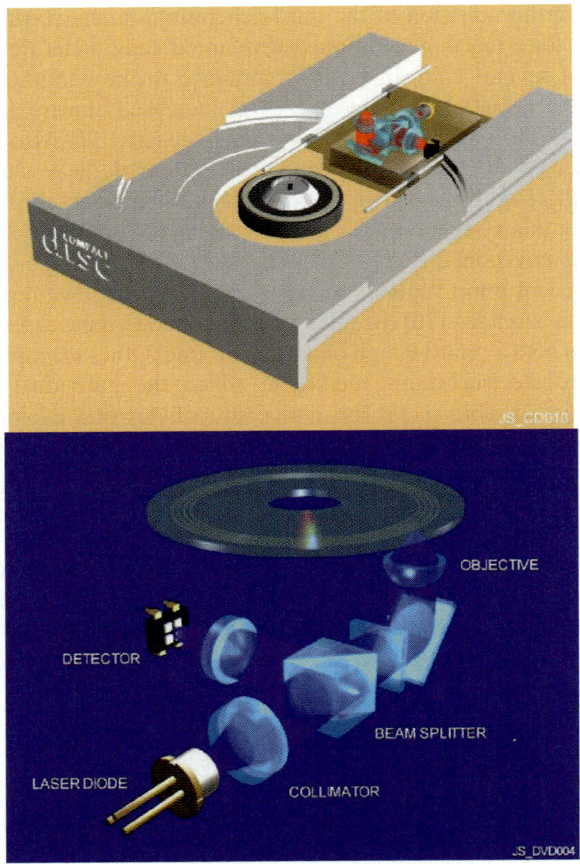

Figure 1. Schematic drawings of an optical storage system. The upper image represents an optical drive to house the disc. The lower image shows a schematic of the optical path with laser, optics, detector and objective lens.

1.2.2.   Basic principles of optical data storage

The principle of optical data storage and retrieval is explained in Figure 2 in a simplified form. The audio and video signals perceived by users are of analogue nature. It is, however, more convenient and robust to use the digital domain to efficiently store, transmit, and retrieve such signals. For this purpose analogue-to-digital conversion (A/D conversion) is done and additional data bits facilitating error correction (error correction coding, ECC) are added. In its digital form the user data is a binary code represented by a sequence of bits defined as logical "1"s and "0"s.

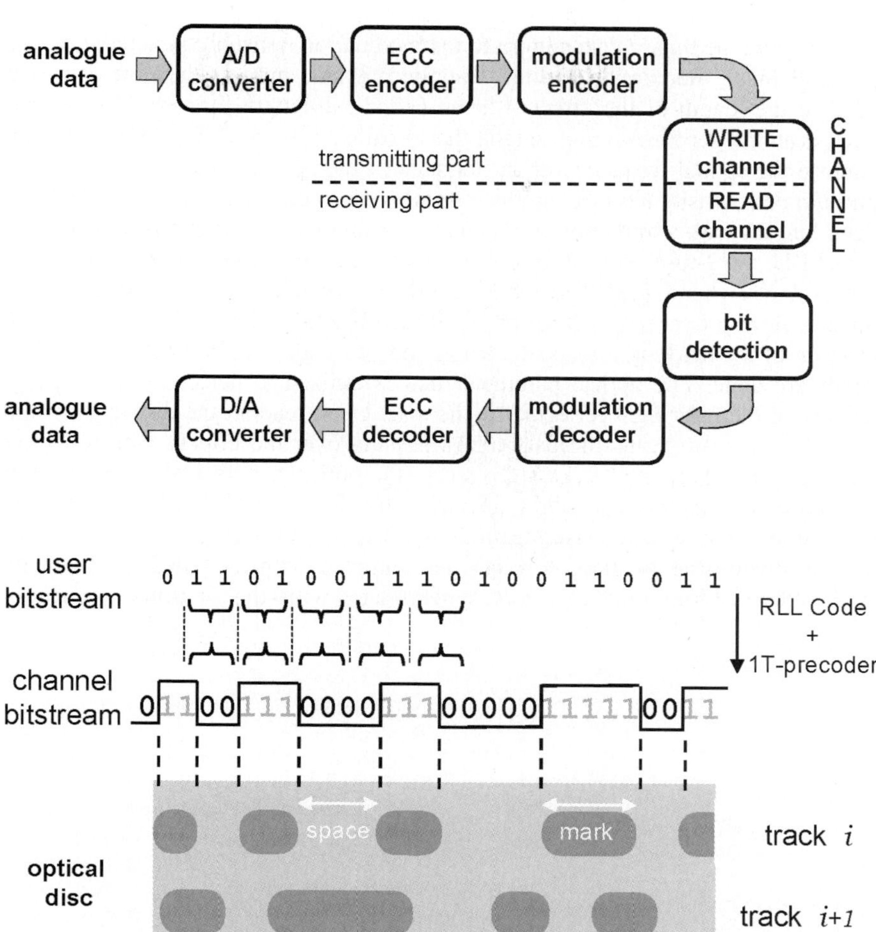

Figure 2. Principles of optical data storage. The upper panel represents a schematic flow chart of the information flow in an optical storage device. The lower panel denotes a typical data pattern in an optical disc and the corresponding channel and user bit stream.

In optical discs, data is represented by small areas (marks or pits) with optical properties that are different from the optical properties of the surrounding matrix. Marks (pits) and spaces (lands) between them are often referred to as (marking) effects. Typically, an optical medium is designed such that the reflectivity of marks (pits) is lower than the reflectivity of the surrounding matrix at the laser wavelength used. In case of recordable and rewritable media, the written areas (marks) have an intrinsically different reflectivity upon thermal degradation. Amplitude modulation is the main mechanism for readout of data. ROM media are mass-replicated and are in most cases provided with a metallic mirror. Constructive and destructive interference of the focused laser spot causes modulation, also referred to as phase modulation. To adapt the binary data pattern to the modulation transfer characteristics of the optical channel, modulation coding is applied. In this process the user data is encoded in the length of the effects (the so-called run-length limited, RLL, coding), which is an integer times a unit-length, the so-called bit-length. To obtain an optimal match to the spatial frequency characteristics of the optical channel and to achieve optimum data density a set of lengths is employed. In the case of CD and DVD a set of run lengths with a minimum of 3 and maximum of 11 channel bits is used. In the case of BD, the 2-to-8 set is used. More details on encoding and error correction can be found elsewhere. [13] The marking effects are placed in data tracks, which typically form a concentric spiral on the disc substrate. To retrieve the information detection of the marking effects, decoding and subsequent conversion into analogue signals are done. The optical parameter that is utilized to detect the effects is the intensity of the laser light reflected by the disc. Upon readout the disc spins and the focused laser beam scans the data tracks passing over the effects. The reflectivity level difference between marks and spaces (the optical contrast of the effects) and the interference in the laser light diffracted by the effects pattern yield intensity modulation of the reflected laser beam. In order to establish the lengths of the effects the intensity profile is sliced through and sampled with a predefined frequency, which is derived from (and, therefore, synchronized with) the rotational frequency of the disc.

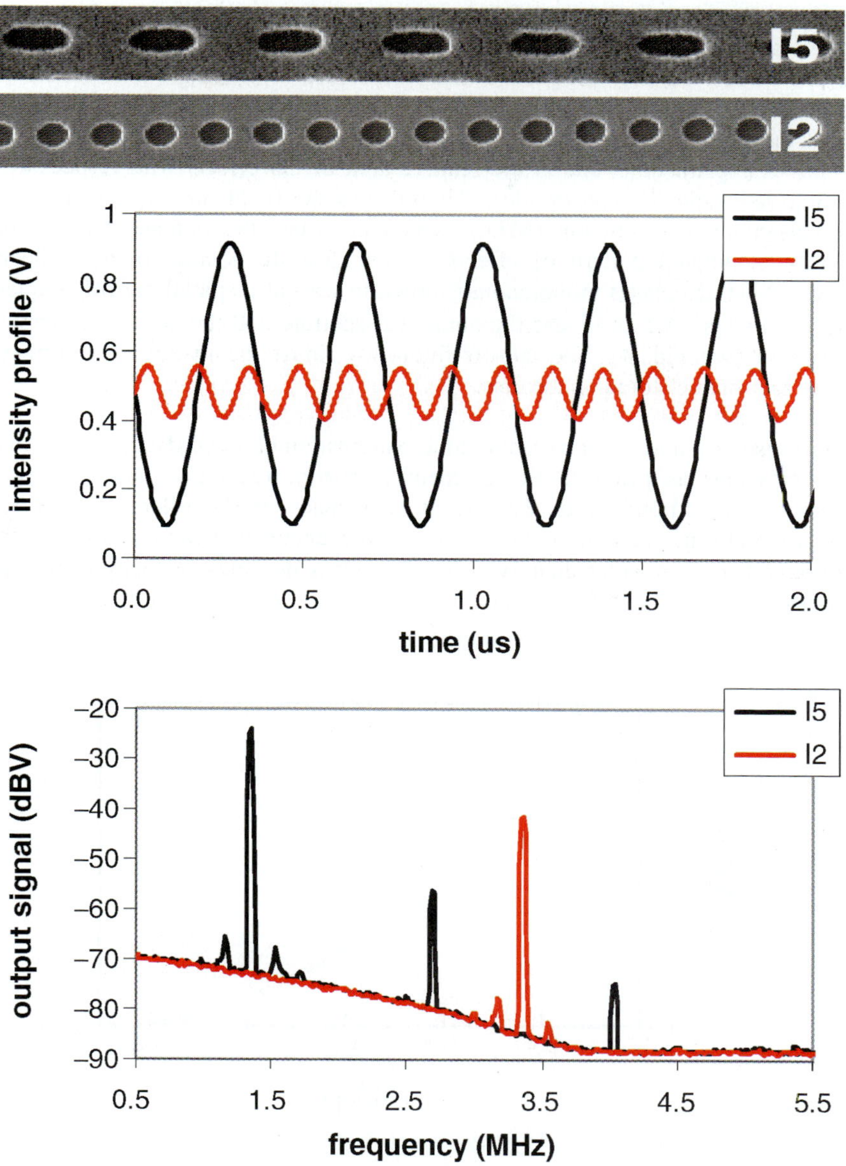

Figure 3. Time (upper plot) and frequency (lower plot) domain signals from single-tone data carriers (I2 refers to a 2T single tone; I5 refers to a 5T single tone).

To give a simple example, modulation profiles of two different single-tone data patterns are plotted in time and frequency domains in Figure 3. An important fact that can be derived from the plots is that in the case of the single-tone carrier with shorter effects (higher frequency) the modulation amplitude is smaller compared to that of the single-tone carrier with longer effects (lower frequency). There are two reasons to explain this. One is the relative area of the effects with respect to the effective laser spot size on the disc. The other is the frequency dependence of the modulation transfer function (MTF), which describes the optical response of a spatially modulated pattern of effects on the disc. In central-aperture-detection systems, MTF decreases monotonously down to zero at a spatial cut-off frequency $cf=2NA/\lambda$, where NA and λ are the numerical aperture and the laser wavelength of the system (see Fig. 4). The cut-off frequency limits the maximum information density that can be stored on the disc.

In the frequency domain such single-tone patterns (single-tone data carriers) manifest themselves as peaks at the frequencies (main frequency plus higher-order harmonics) corresponding to the spatial-frequencies of the effects on the disc. Translated into the signal domain, the signal frequencies that can be extracted from an optical disc are smaller than $2\nu NA/\lambda$, where ν is the linear velocity of the spinning disc.

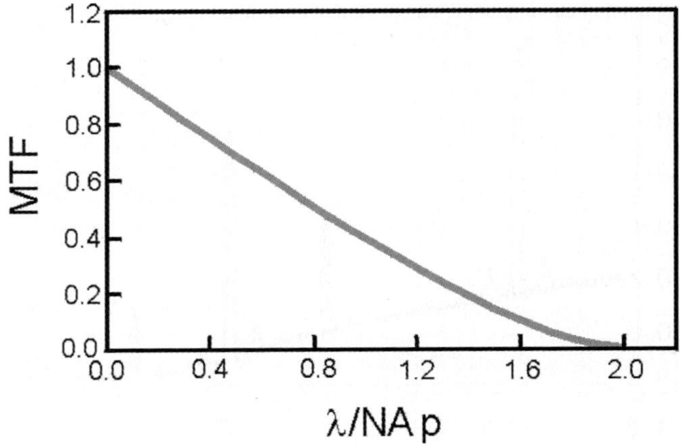

Figure 4. Modulation transfer function of a central-aperture optical channel.

As may be obvious from the above, bit detection is directly related to accurate measurement of the intervals between the slicer crossings in the time domain. Any deviation in lengths of the effects, irregularities in their shape or in local disc reflectivity, cross talk with the neighboring tracks, as well as fluctuations in electronics and laser performance etc. will inevitably alter the intensity modulation profile and affect detection. In panels (a) and (b) of Figure 5, two intensity modulation profiles obtained for a random sequence of bits are shown. These intensity modulation plots are called the eye-patterns. The eye-pattern in panel (a) corresponds to a perfect case. The eye-pattern in panel (b) corresponds to a case where imperfections are present. As can be seen from the figure the presence of imperfections causes spread in the intensity modulation profiles. When the sources of fluctuations are Gaussian in character, the standard deviation of the Gaussian time distribution is called *jitter* and is expressed as percentage of the clock-time: *jitter*$=\Delta t/2T\times100\%$, where Δt is the spread at the slicer-level crossings and T is the time-period. Each mark and space (pit and land) length can be defined as its average length in time domain and jitter in percent of clock-time. An increase in jitter manifests itself in the time-frequency domain as a decrease in the signal strengths, which is characterized by the signal-to-noise ratio (SNR). The relation between jitter and SNR can be expressed as *jitter*$=\frac{1}{2}\times10^{-SNR/20}\times100\%$.

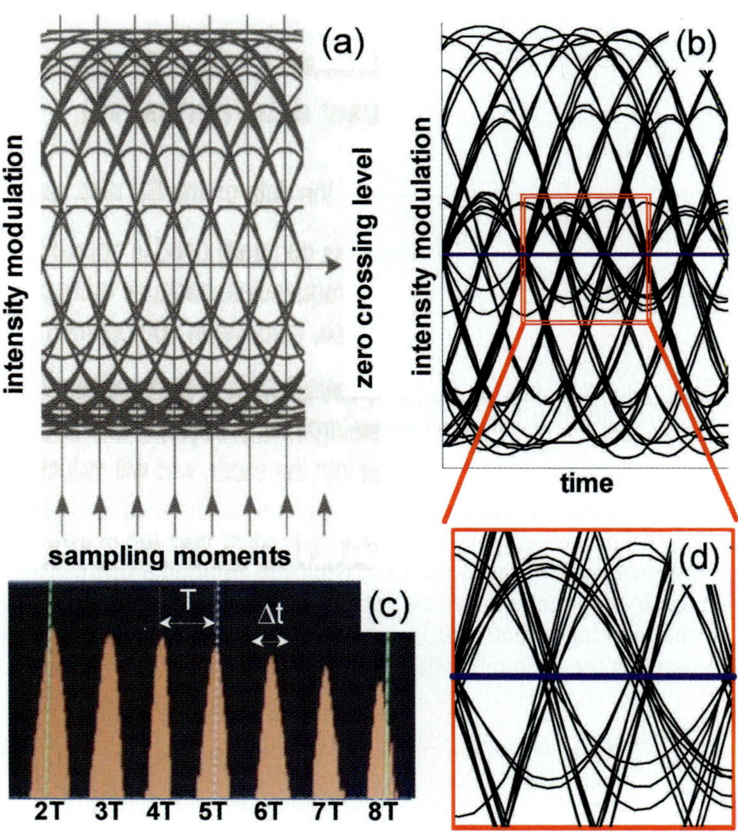

Figure 5. a – eye-pattern calculated for a perfect case; b – eye-pattern with imperfections included; c – measured jitter histogram for a 2T. 8T data pattern; d – magnified section of panel (b).

When analyzing recording media the concept of carrier-to-noise ratio (CNR) has proven to be useful. This CNR is the SNR of a single-tone data carrier written on the disc. By contrast to normal SNR, CNR is measured in a narrow bandwidth centered at the carrier frequency.

In turn, full bandwidth SNR is related to the bit error rate (BER), which is ultimately a figure of merit for the quality of data storage and retrieval. The relationship between BER and SNR for a threshold detection system is given in Figure 6, which displays a pronounced increase in BER with decreasing SNR. The science behind this graph can be found elsewhere. [14]

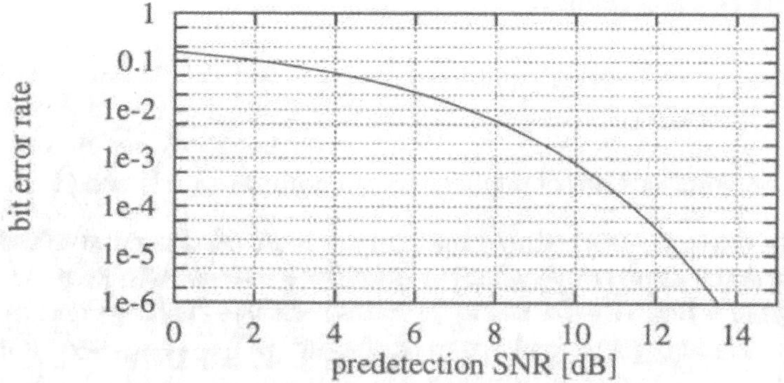

Figure 6. Dependence of bit error rate on signal to noise ratio over total bandwidth, taken from [14].

To facilitate bit detection, *equalization* is typically employed in optical storage systems. [15] An improvement is achieved by electronically boosting the high-frequency response and, in this way, increasing the amplitude of intensity modulation generated by the smaller effects. On the media side, enhancing the optical contrast of the marking effects can increase the modulation amplitude. This aspect will be discussed in the upcoming chapters.

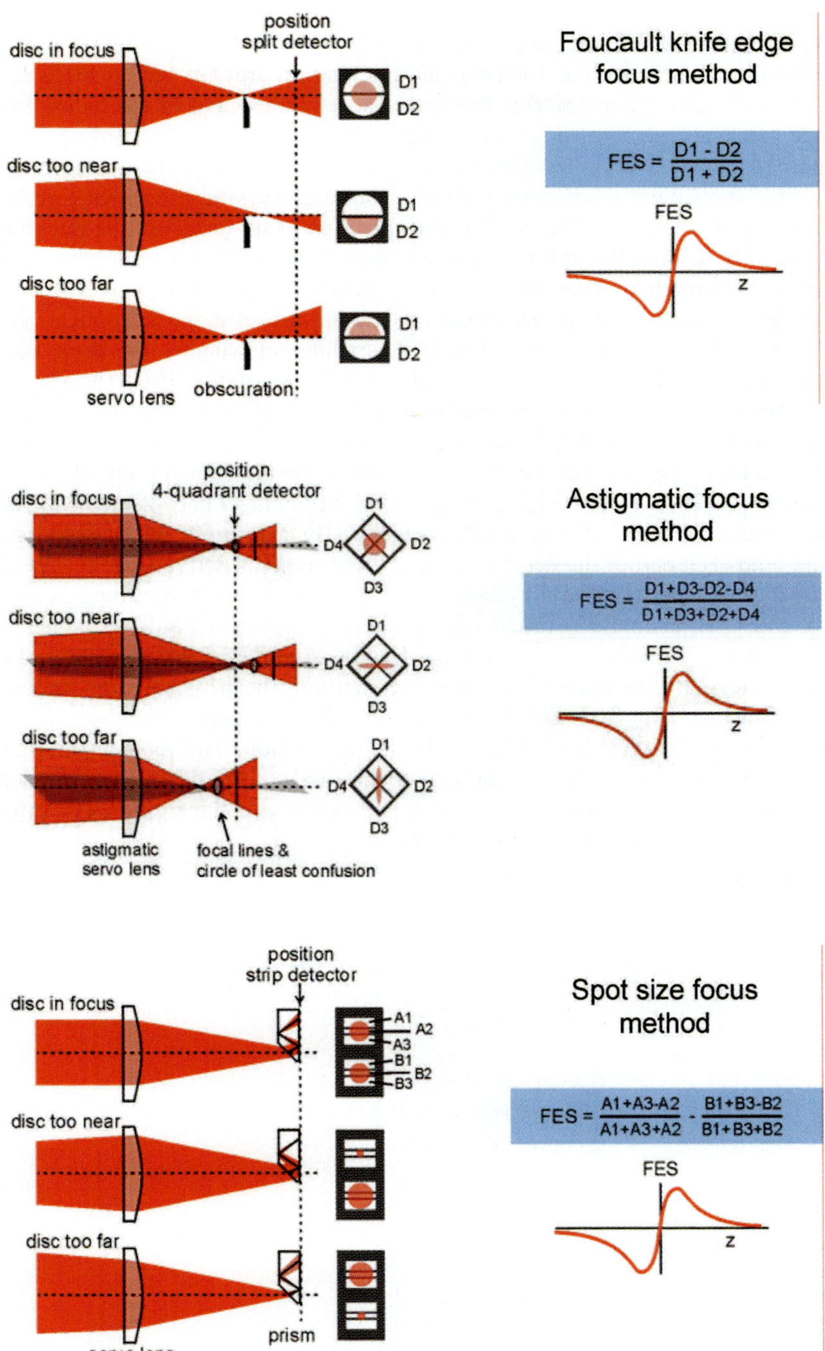

Figure 7. Focusing methods.

To realize accurate bit detection a number of functions of the drive have to be well under control. These include focusing and tracking. In order to stay in-focus and on-track a continuous adjustment of the lens-disc separation and of the radial position on the track are performed by the drive during read-out and recording. For this purpose the lens is mounted into an actuator, which allows electro-mechanically controlled movement of the lens. The focusing and tracking processes consist of a dynamic measurement of the amount and direction of de-focus and de-tracking and subsequently feeding this information into the actuator to do the appropriate corrections. Several methods exist to accomplish dynamic measurement of de-focus, see Figure 7. All these methods are based on making use of a special optical element that shapes the beam in a certain way depending on whether the laser beam is focused in front, behind or right onto the data layer of the disc. The element is complemented with a dedicated photo-detector. The element and the detector are placed into the laser beam reflected by the disc. In the case of the Foucault focusing method, a knife is positioned on the optical axis at the ideal focal point of the returning beam. Depending on the focus position the knife cuts a part of the beam, which is subsequently projected onto a split detector. By measuring the amount of light falling onto each part of the detector a focus error signal is derived. In the case of the astigmatic focusing method a cylindrical lens is placed in the returning beam. The lens creates perpendicularly oriented astigmatic lines on either side of the best focal point position. A quadrant detector is used to measure the relative intensity of these lines. In the spot-size focusing method the returning beam is split in two using a wedge. The two beams form two spots on the photo-detector. The size of the two spots mutually changes depending on the focus position. The focus error signal is derived from the relative size of the two spots. In all of the three cases the derived error signal has an S-shape. The intensity and polarity of the signal carry information on the amount and direction of defocus.

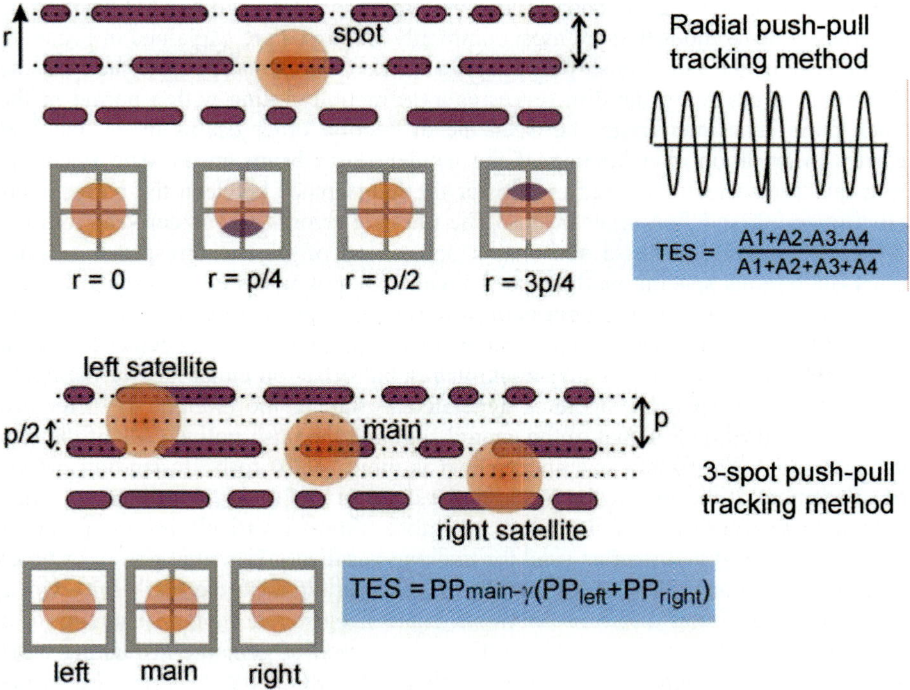

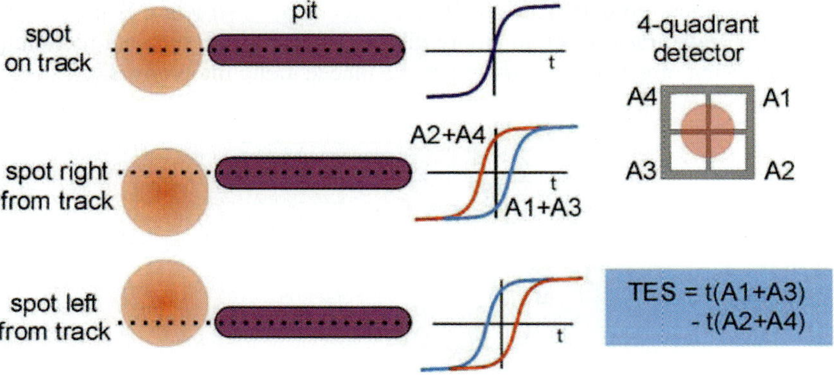

Differential phase detection tracking method

Figure 8. Tracking methods.

There also exist several methods for dynamic measurements of the radial position on the track. The methods that are most commonly employed are explained in Figure 8. In the case of the radial push-pull tracking method use is made of the fact that the data track structure on the disc serves as a diffraction grating with a period of the track pitch. The interference between the diffraction order beams in the far-field carries information over landing of the incident laser beam on the data tracks. A four-quadrant detector is used to register the interference between the partly over-lapping zeroth and first-order beams. The tracking error signal is derived from the difference signal and has a sine-shape, one period of which corresponds to one-track-pitch radial spacing on the disc. To realize 3-spot push-pull tracking a diffraction grating is placed in the light path of the drive to generate satellite beams. The whole setup is arranged such that when the main beam falls onto the center of a track the satellite beams land with a ½-trackpitch radial offset on either side of the track. The radial tracking error signal is generated by taking the (weighted) difference between the push-pull signal of the central spot and the push-pull signals of the two satellite spots. The 3-spot push-pull signal is more robust to beam landing offsets (displacement of the spot with respect to the detector due to e.g. misalignment) than the single spot push-pull signal and is, therefore, almost invariably used in practice. One more method that is often used for tracking is called differential phase (or time) detection, DPD or DTD. If a diffraction-limited spot lands onto a mark (pit) with a radial offset a timing difference between signals registered by the quadrants of a four-quadrant detector occurs. This difference is used to generate a tracking error signal. This error signal is particularly suitable for ROM-discs, where the marks needed to derive the signal are always present.

In the case of pre-recorded discs, the presence of marks (pits) and spaces (lands) is sufficient to generate the radial tracking error signal. In the case of recordable and rewritable discs where no data is originally present a groove structure is introduced into the disc to make tracking of an empty disc possible. During the data recording process the marking effects representing data are placed along the grooves.

1.2.3.   Optical storage roadmap

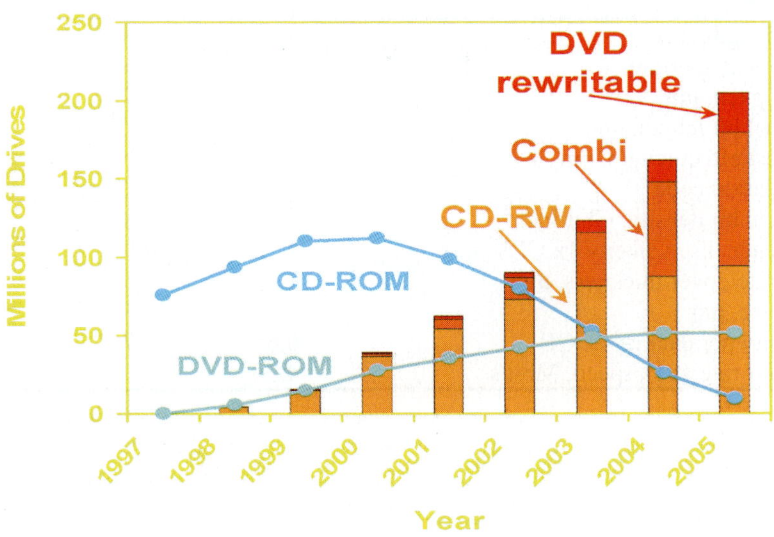

Figure 9. Optical disc storage technologies roadmap.

The technology roadmap in optical storage is usually characterized by the disc capacity and data transfer rate. The overall trend is shown in Figure 9. The raise in storage capacity is achieved through increase in storage density (channel bit length and track pitch), number of data layers, and the efficiency of coding schemes and signal processing. Typical parameters, which characterize the trend are presented in Table 1. The density increase is realized by employing lasers with shorter wavelengths and objective lenses with a higher numerical aperture. Aided by coding efficiency, the storage densities that have been achieved in CD, DVD, and BD are 0.4 Gbit/inch^2, 2.8 Gbit/inch^2, and 14.7 Gbit/inch^2, respectively. The maximum velocity of the spinning disc limits data transfer rates. At 1x speed the transfer rates amount to 0.49 Mbps (CD), 11 Mbps (DVD), and 36 Mbps (BD) with the maximum of 56x, 16x, and 12x for the three systems, respectively. This maximum data transfer rate is dictated by the servo characteristics of current optical drives rather than recording material or disc/substrate characteristics.

Table 1. Characteristic parameters of CD, DVD, and BD systems.

Parameter	CD	DVD	BD
Wavelength, nm	780	650	405
Numerical aperture	0.45/0.5	0.60/0.65	0.85
Track pitch, μm	1.6	0.74	0.32
Channel bit length, nm	277	133	74.5
Shortest effect length, nm	831	399	149
Modulation code	EFM	EFM+	17PP
Physical bit density, Gbit/inch^2	0.4	2.8	14.7
Reference disc velocity 1x CLV, m/s	1.2	4.0	4.92
Substrate/cover thickness, mm	1.2	0.6	0.1
Spot size, μm	0.9	0.55	0.238
Capacity per data layer, GB	0.65	4.7	25.0
Transfer rate at 1x speed, Mbit/s	4.3	11	36

1.2.4.   Optical media

An optical medium (often referred to as optical data carrier) typically comprises a disc-shaped substrate, one or more data layers, and a dummy substrate or a cover. Often, discs are complemented with labels carrying user information such as a table of contents of the data stored on the disc, etc. A cross-sectional view of a dual-layer DVD disc is given in Figure 10. The laser beam accesses the data layers through the bulk of a transparent material. One of the major advantages of such a media configuration is that the data layer is well protected from potential damage caused by disc handling. Typical defects such as scratches, fingerprints, dust etc. present on the disc surface are far out of focus of the addressing laser beam, and therefore hardly hamper the quality of the readout signals. In this way, the overall system robustness is greatly improved in comparison to direct contact systems, such as the vinyl LP-disc system and makes a cartridge kind of protection system redundant (like in hard disk drives or magnetic tape systems).

The technological choice of decreasing the laser wavelength and increasing NA of the objective lens in order to improve storage density comes at the cost of operating margins, such as disc tilt and focus error. In order to keep the margins at an acceptable level the thickness of the transparent material through which the data layer is accessed has to be reduced from 1.2 mm in the case of CD to 0.6 mm in the case of DVD to 0.1 mm in the case of BD. To facilitate backwards compatibility through the whole range of optical discs the total disc thickness needs to be kept at 1.2 mm. Thus, a CD is recorded and readout through the disc substrate whereas a BD is accessed through a 0.1 mm thick cover, which is brought onto a 1.1 mm thick (dummy) substrate. A DVD comprises of two 0.6 mm thick substrates bonded back-to-back. The disc substrate is usually made of polycarbonate. This material is relatively easy to process via injection molding, it is transparent at the utilized laser wavelengths, and it is inexpensive and has a low moisture absorption resulting in a more stable shape. The cover layer is typically made of a polycarbonate sheet or a layer of resin.

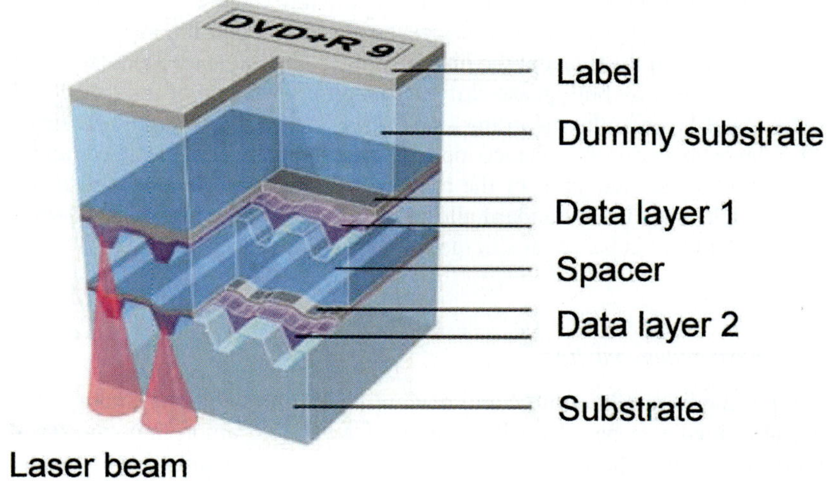

Figure 10. A cross-sectional view of a dual-layer recordable DVD disc.

With respect to application, optical storage media can be divided in three types. The media types are usually being referred to as read-only (ROM), recordable or write-once (R or WO), and rewritable (RW or RE). The physical difference between these media is in the type and structure of the data layer.

The ROM media can only be read-out but cannot be erased or recorded. The data layer of such media contains pits replicated in the substrate during the disc manufacturing process. The manufacturing processes that are most widely employed are injection molding and photo-polymerization. The relief structure of the pits and lands is complemented with a thin reflective layer to facilitate readout of the data. Upon reflection from the pit and the land areas, rays of the laser beam gain a phase shift. Interference between the light rays results in modulation of the intensity of the reflected light. The depth and the width of the pits are chosen such that the intensity modulation is optimal.

In the case of R/WO/RW/RE media the marking effects representing the data are small recording marks formed in the recording data layer by the laser beam of the optical drive. The R/WO media can be written only once but read-out many times.

Many recording mechanisms and materials systems have been proposed to realize R/WO media. These include hole burning, alloying of bi-layers, altering surface or interface roughness, agglomeration (island formation) in thin films, altering material state/phase, or bleaching. At present, most of the recordable CD and DVD discs are based on organic dyes. A layer of dye is typically brought onto the disc by a spin-coating process and is a part of a recording stack, which also comprises a metal and a dielectric layers. During recording the dye is locally heated and degraded (bleached) with a focused laser beam in an irreversible manner. The degradation is accompanied by a change in optical properties of the dye and the local geometry of the interface between the dye and the disc substrate. The intensity modulation during read-out is generated by both phase shift and amplitude change of the reflected laser light. The metal layer in the recording stack serves as a reflector and heat sink. The dielectric layer is used to enhance optical contrast between bleached and non-bleached areas of the dye, and for the purpose of chemical and mechanical protec-tion in the stack. The BD-R standard allows also for inorganic material systems, like the Cu-Si system that is based on silicide formation upon laser heating.

1.2.5.   Phase-change media

The RW/RE media can be written and readout many times. The technology utilized in rewritable media is based on laser-induced reversible amorphous-to-crystalline transitions in a thin phase-change film. The amorphous marks have typically a different reflection than their crystalline surrounding. The difference in reflection results in optical contrast that enables the readout of data. The readout principle is schematically illustrated in Figure 11. In the top panel a data pattern of amorphous marks in the crystalline matrix is shown, which is in this case visualized by Trans-mission Electron Microscopy (TEM). In the bottom panel a reflectivity profile corresponding to this data pattern is sketched. The amorphous marks result in a drop in the reflectivity level, which is detected as signal modulation.

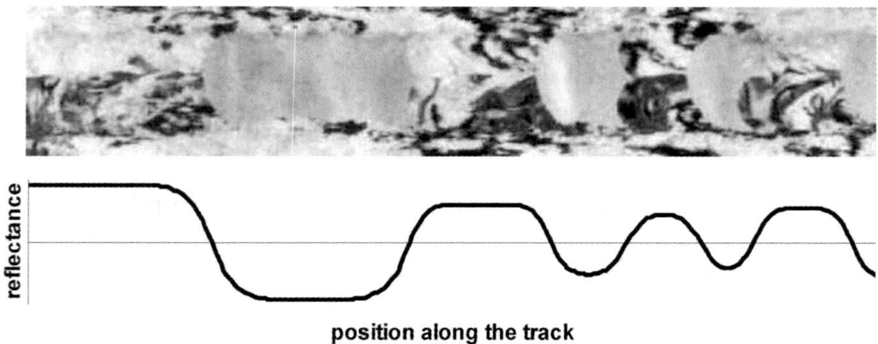

Figure 11. Schematic of the readout principle of amorphous marks in a crystalline layer.

The phase-change layer is a part of the recording stack. It is sandwiched between so-called interference layers. A metallic layer is added to the stack on the side opposite to the entry side of the laser beam. A basic recording stack structure is sketched in Figure 12. For convenience, stacks are often denoted with a series of letters, MI_2PI_1 in the case considered here, where M stands for metal layer, I stands for interference layer, P stands for phase-change layer and so forth. The indices indicate the layer order in which the incident laser beam penetrates the stack.

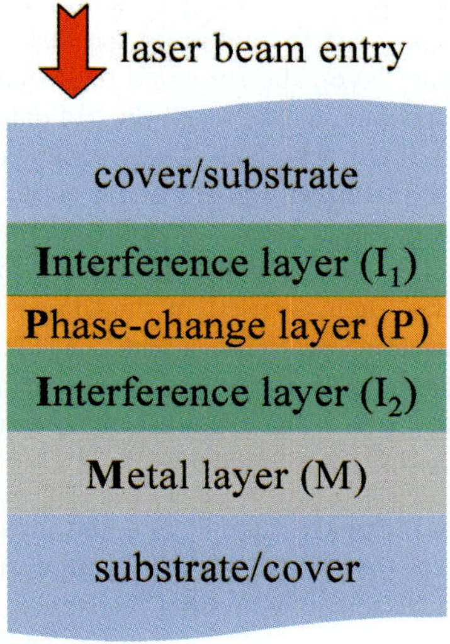

Figure 12. Schematic of a phase-change recording stack.

All layers in the stack fulfill multiple functions. The phase-change layer acts as a signal modulation enabler and as a medium where data can be stored and erased. The metal layer works as a reflector and a heat sink. The interference layers serve for optical contrast enhancement, thermal resistance, and mechanical and chemical protection. Additional layers are often used to promote material crystallization, to improve the mechanical or chemical stability, etc. The thickness of the layers and their composition is of utmost importance for the recording stack performance.

Phase-change compositions that are used for rewritable optical discs are discussed in chapter 2. The high absorption coefficient and relative low thermal conductivity of these materials hamper mark formation (melting) in too thick phase-change layers. A too thin layer will not provide sufficient contrast between the amorphous and crystalline state, preventing the accurate detection of marks and decoding of data. Furthermore, a very thin phase-change layer may possess a low chemical stability.

The optimum thickness of the phase-change layer, typically between 5 and 30 nm, depends very much on the application. It is a compromise between good optical contrast, excellent recording properties and sufficient chemical stability.

A phase-change recording stack usually comprises a metallic layer for two important reasons. In the first place, the metals used possess a high absorption coefficient and a low index of refraction. In combination with the other layers in the recording stack, this leads to a high stack-reflection and an improved readout of the amorphous marks (improved modulation). In addition, the metals are used to improve the thermal response during writing and erasing of amorphous marks. Metals have a high thermal conductivity, which is favorable for the fast heat removal after melting of marks in the phase-change layer, the so-called melt-quenching process. Also for direct overwrite of the amorphous marks, when the old data need to be removed in a single passage during write of the new data, it is advantageous that the old marks are completely erased by heat diffusion ahead of the write pulse. Suitable metallic materials are alloys based on Ag, Al or Au and generally comprise a dopant to improve the chemical stability (for instance to control the grain size). For semi-transparent recording stacks, such as used in dual-layer phase-change discs, thin metallic layer or semitransparent heat sink layers, i.e. ITO, Al_2O_3 or HfN, can be applied to guarantee sufficient cooling rate and sufficient transmission to access the second recording stack as well. The application of these materials and their recording characteristics are discussed in chapter 5.

The dielectric film between the phase-change film and the metallic heat sink layer is primarily required to control the heat diffusion through the recording stack during erasing and writing of data. It acts as a thermal resistance for the heat flow into the metallic layer. In addition, the dielectric layers impose a stable chemical barrier to prevent diffusion of components out of the phase-change film. The dielectric layers contribute also to optimum optical stack characteristics. ZnS-SiO_2 is commonly used as dielectric interface material in a phase-change recording stack. It has a low thermal conductivity, it is optically transparent from 400 nm to 800 nm (thus for CD, DVD and BD applications), it has a relatively high index of refraction and it is thermally stable. A lot more materials have been considered for application in optical discs, such as HfN, Al_2O_3, and ITO but also SiC, Si_3N_4, TiO_2, SiO_2, etc. Of course, the applicability of these materials depends, among others, on the wavelength of the used laser light and the optical characteristics of the materials. The upper dielectric layer is primarily used to optimize the optical contrast of the recording stack. Also the high-temperature-resistant dielectric layer acts as a thermal barrier towards the substrate (CD, DVD) or cover layer (BD). ZnS-SiO_2 is also the preferred material.

1.3.   *Scope of this book*

The main purpose of the book is to provide the reader with a detailed overview of the basics behind optical phase-change recording. Although the emphasis will be mainly on the material aspects of optical phase-change recording, in many cases it is inevitable to discuss hardware and signal processing details.

The layout of the book is as follows. Theoretical aspects of phase-change materials are dealt with in Chapter 2. In Chapter 3, the thermal modeling of phase-change recording is described and main characterization techniques and methodologies are explained. Chapter 4 gives an extensive analysis of the data storage process in rewritable phase-change media. Two main applications areas, namely high-speed and dual-layer recording are addressed in Chapter 5.

1.4.   References Chapter 1

[1]   K. Compaan, P. Kramer: Philips Tech. Rev. 33, 178 (1973)

[2]   http://www.ecma-international.org

[3]   Recordable compact disc system description, Part III: CD-RW, version 2.0, Aug 1998, Royal Philips Electronics. Recordable compact disc system description, Part III: CD-RW, volume 2: high speed, version 1.1, June 2001, Royal Philips Electronics. Recordable compact disc system description, Part III: CD-RW, volume 3: ultra speed, version 1.1, July 2003, Royal Philips Electronics.

[4]   DVD+RW 4.7 Gbytes Basic Format Specifications, volume 1, version 1.3, September 2004, Royal Philips Electronics. DVD+RW 4.7 Gbytes Basic Format Specifications, volume 2, version 1.0, December 2004, Royal Philips Electronics.

[5]   M. Kuijper, I. Ubbens, L. Spruijt, J. M. ter Meulen and K. Schep: Proc. SPIE 4342 (2001) 178, T. Narahara, S. Kobayashi, M. Hattori, Y. Shimpuku, G. van den Enden, J. Kahlman, M. van Dijk and R. van Woudenberg, 2000, Optical disc system for digital video recording, Jpn. J. Appl. Phys., Vol. 39 Part 1, No 2B, pp. 912-919.

[6]   K. Schep, B. Stek, R. van Woudenberg, M. Blum, S. Kobayashi, T Narahara, T. Yamagami and H. Ogawa, 2001, Format description and evaluation of the 22.5 GB DVR disc, Jpn. J. Appl. Phys., Vol. 40. and M.J. Dekker, N. Pfeffer, M. Kuijper, I.P.D. Ubbens, W.M.J. Coene, E.R.Meinders, and H.J. Borg, 2000, Blue phase-change recording at high data-density and data rates, SPIE 4090, pp. 28-35.

[7]   C. A. Verschuren, J. M. A. van den Eerenbeemd, F. Zijp, Ju-Il Lee, D. M. Bruls, Near-Field Recording with a Solid Immersion Lens on Polymer Cover-layer Protected Discs, Jpn.J.Appl.Phys. 45, No. 2B, pp. 1325. C. A. Verschuren, F. Zijp, J. M. A. van den Eerenbeemd, M. B. van der Mark and Ju-Il Lee, Towards Cover-Layer Incident Read-Out of a Dual-layer Disc with a NA =1.5 Solid Immersion Lens, Japanese Journal of Applied Physics, Vol. 44, No. 5B, 2005, pp. 3554–3558.

[8]   Padiy A, Yin B, Verschuren C, et al., Signal processing for 35GB on a single-layer Blu-ray disc, SPIE Proceedings, Optical Data Storage, vol. 5380, pp. 56-70, 2004.

[9]   D.M. Bruls, A.H.J. Immink, A.M. van der Lee, W.M.J. Coene, J. Riani, S.J.L. van Beneden, M. Ciacci, J.W.M. Bergmans and M. Furuki, "Two-Dimensional Optical Storage: High-speed read-out of a 50 GByte single-layer optical disc with a 2D format using lambda= 405nm and NA = 0.85", Japanese Journal of Applied Physics,Vol. 44, No. 5B, 2005, pp. 3547-3553.

[10]   Holographic Data Storage, H.J. Coufal, D. Psaltis, G.T. Sincerebox (eds.), Spinger Verlag, Optical Sciences, Berlin (2000).

[11]   Multi-stack information carrier based on electrochromic switching, WO2004077422 A1, 2004, WO2004077410 A2, 2004, WO2004077414 A1, 2004.

[12]   M. Mansuripur and P. Khulbe, 'Macromolecular data storage with petabyte/cm3 density, highly parallel read/write operations, and genuine 3D storage capability (invited paper), Optical Data Storage Conference, Monterey, California, April 2004.

[13]   K. Schouwhamer-Immink: Coding Techniques for Digital Recorders, Prentice Hall, Englewoods Cliffs, NJ 1991

[14]   J.W.M. Bergmans: Digital Baseband Transmission and Recording, Kluwer Academic Publishers, Boston/London/Dordrecht, 1996, IBSN nr 0-7923-9775-4.

[15]   E.F. Stikvoort, J.A.C. van Rens, *IEEE J. On Selected Areas in Communications* **10** (1992) 191.

2.   Theoretical aspects of phase-change alloys

2.1.   *Introduction*

As mentioned in the previous chapter, in a rewritable disc information is stored in the so-called phase-change layer. [16], [17], [18] This is often a chalcogenide alloy, which can be reversibly converted from the crystalline to the amorphous state by a laser pulse. Applying short consecutive write pulses that cause melt quenching of the initially crystalline recording film controls writing of amorphous marks. Erasure of amorphous marks is enabled by heating the phase-change film to intermediate temperature levels to induce re-crystallization. Characteristic temperature-time profiles that are associated with the write and erase processes are given in Figure 13. Typically, the short high-power write pulse leads to a steep temperature rise and subsequent sharp fall after switching off the laser. Due to the short duration of the write pulse no significant heat dissipation occurs and the rapid temperature rise induces local melting of the material. After switching off the laser power, the temperature drops very quickly to below the crystallization temperature of the material. The temperature drop leads to a drastic reduction of atomic mobility, which is negligible at room temperature. Because the cooling-down period is very short, atoms do not have enough time to return to the stable crystalline state and become trapped in a metastable, amorphous state. In this way small amorphous marks are created in the crystalline matrix of the phase-change layer. Due to the difference in the optical properties of the amorphous and crystalline state, information can be derived from a change in the reflectance. The recorded information can be erased by heating the material with the same focused laser beam above its crystallization temperature but below the melting temperature. The crystallization temperature is often close to the glass transition temperature T_g denoted in Figures 2 and 3. Since the atoms are very mobile at elevated temperatures, the amorphous state rapidly returns to the crystalline state. Thus, erasure is controlled re-crystallization of amorphous marks induced by a moderate temperature rise during a period of time that is long enough with respect to the typical crystallization time of the phase-change material.

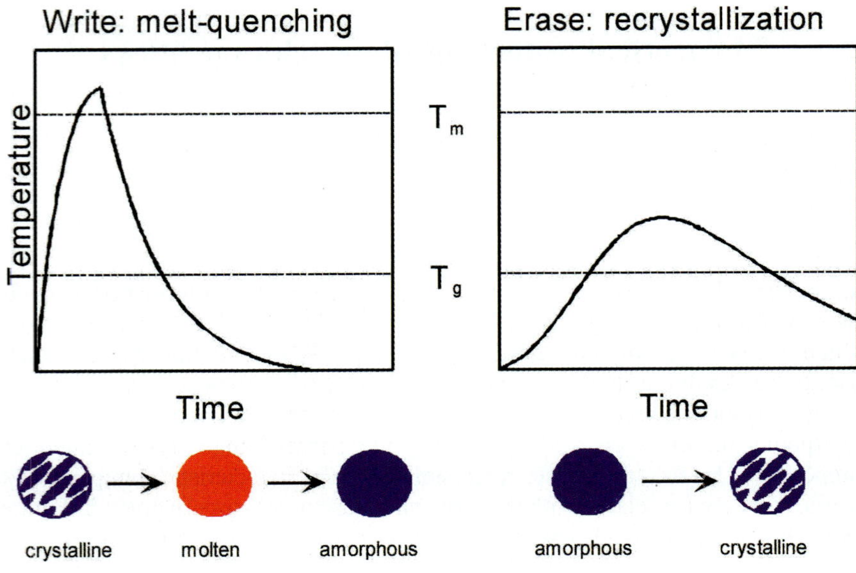

Figure 13. Schematic of the temperature-time profiles associated with recording (left panel) and erasure (right panel) of amorphous marks in a crystalline layer.

While a majority of materials can be amorphized if cooled rapidly enough, just a few materials show a pronounced difference in optical properties between the amorphous and crystalline states. [18] The presently used phase-change materials are a result of a 30-year and still continuing period of empirical optimization of materials. Table 2 shows a historical overview of the materials development for phase-change media. [18] A large number of phase-change materials have been proposed, but only a few materials meet all requirements. Therefore, it is worthwhile to discuss the requirements of phase-change materials. Rewritable storage media have to fulfill five main data storage requirements. [16] They have to enable writing of data (writability). The stored information has to be stable (archiving) and easy to read (readability). Then the information should also be erasable (erasability) and the storage medium should allow numerous write/erase cycles (cyclability). These data storage requirements can be translated to media requirements (see Table 3).

Table 2. History of materials development for phase change media (adapted from [18]).

Year	Composition
1971	Te-Ge-Sb-S
1974	Te-Ge-As
1983	Te-Ge-Sn-O
1985	Tn-Sn-Se, Ge-Se-Ga
1986	Te-Ge-Sn-Au, Sb_2Se, In-Se, GeTe, Bi-Se-Sb, Pd-Te-Ge-Sn
1987	$GeTe$-Sb_2Te_3,($Ge_2Sb_2Te_5$, $GeSb_2Te_4$), In-Se-Tl-Co
1988	In-Sb-Te, In_3SbTe_2
1989	$GeTe$-Sb_2Te_3-Sb, Ge-Sb-Te-Pd, Ge-Sb-Te-Co, Sb_2Te_3-Bi_2Se_3
1991	Ag-In-Sb-Te
2001	Ge-In-Sb-Te
2004	Ge-In-Sn-Sb

In this chapter the different requirements that any potential phase-change material should fulfill are discussed. Writability implies easy formation of amorphous marks, i.e. that the material can form a glass upon irradiation with (ns) laser pulses. We will start in section 2.2 with the basics of glass formation and explain what an amorphous solid looks like. For the stored information to be stable, we need to have a stable amorphous state, which requires high activation energy for re-crystallization. To give an impression, amorphous bits should be stable for 30-100 years in media (discs) at room temperature. On the other hand, to enable high data-rate recording, fast erasure at elevated temperatures should be possible. This necessitates a better understanding of the crystallization process and means to improve it. Such improvements should lead to an erasure time per bit of less than 10 ns. Such short times are essential for applications where high data transfer rates are mandatory, such as recording of high definition TV signals as well as for computer-based mass storage applications. A (theoretical) introduction to the process of crystallization is given in section 2.3. After the introduction of these important elements in phase-change recording, namely recording of data and erasing them, the phase-change materials that are currently applied in the rewritable versions of CD, DVD and BD are discussed in 2.4. Finally, in section 2.5, emerging directions in phase-change materials research are presented aimed at obtaining a better understanding of atomic arrangements and the origin of the optical and electrical contrast.

Table 3. Requirements for phase change media.

storage requirement	Material requirement	material property
Writability	glass former	melting point/layer design, appropriate optical absorption
archival storage	stable amorphous phase	high activation energy, high crystallization temperature
Readability	large signal to noise ratio	high optical contrast
Erasability	fast re-crystallization	simple crystalline phase, low viscosity
Cyclability	Stable layer stack	low stresses, low melting temperature

2.2.   Glass formation and the amorphous phase

In phase-change optical recording, information is stored by writing amorphous marks in a crystalline phase-change layer. This section will deal with the basics of the process of mark formation and the structure of the resulting amorphous phase. In

Figure 14 the volume of a liquid is considered that is cooled to below the melting temperature. At the melting point T_m, crystallization may occur. As illustrated in Figure 14, crystallization is accompanied by an abrupt change in volume at T_m. It is, however, also possible that the liquid will become 'undercooled', getting more viscous with decreasing temperature, and ultimately a solid phase will form. This solid is called a glass, and the temperature region in which the undercooled liquid acquires the properties of a solid is called the glass transition temperature. [58], [59] Glass formation is characterized by a gradual break in the slope of the volume vs. temperature diagram. For a given composition, the value of the glass transition temperature depends on the cooling rate. This is also illustrated in Figure 14. When cooled slowly (dashed line in Figure 14.), the glass transition temperature will shift to lower values, as the undercooled liquid has more time to adjust its properties to its metastable equilibrium values. However, cooling down slowly also increases the chance for crystallization.

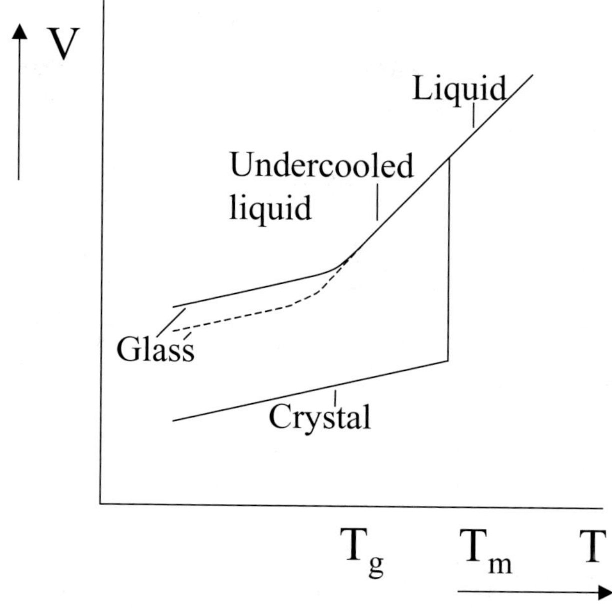

Figure 14. Volume as a function of temperature for a liquid, a glass and a crystal.

While the temperature dependence of the volume of the solid already gives a first idea of the processes prevalent in undercooled liquids, the most relevant property of a glass is the viscosity. [59] This quantity is inversely related with the atomic mobility via the Stokes-Einstein equation. Figure 15 displays the typical temperature dependence of the viscosity of an undercooled liquid. In this metastable state the viscosity increases upon cooling. Below the glass transition temperature the mobility in the material is so low that the metastable equilibrium can no longer be achieved on the accessible time scales. The system is now frozen in the amorphous state (glass). Structural relaxation tends to increase the viscosity of this state and moves it closer to the undercooled liquid. This relaxation proceeds on a very long time scale, however.

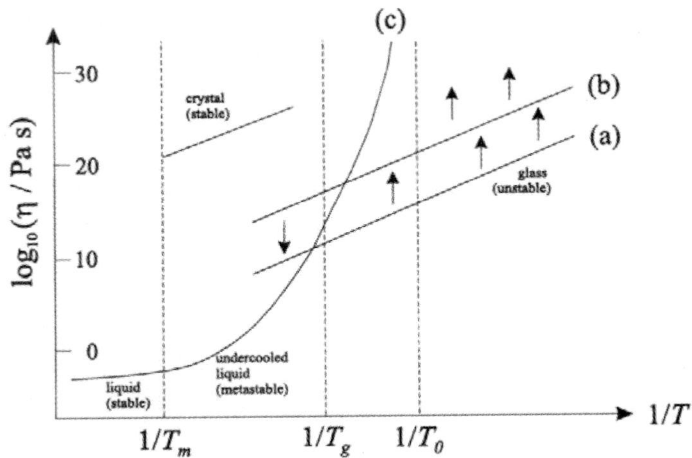

Figure 15. Viscosity of a typical fragile glass. The temperature dependence of the viscosity differs below and above T_g. Below T_g the viscosity shows an Arrhenius-like behavior. Here the glass is in an iso-configurational state, giving rise to a temperature independent activation barrier for self-diffusion. In the undercooled liquid the viscosity follows the Vogel-Fulcher-law. [19], [20] The viscosity is considerably lower than the Arrhenius behavior would predict, because the undercooled liquid is able to adapt the equilibrium configuration suitable for a given temperature. The stronger the temperature dependence of the viscosity in the undercooled liquid the more fragile the glass is [21], [21].

The ability to form glasses is almost a universal property of condensed matter. In order to produce an amorphous material, crystallization should be bypassed. Crystallization takes time and, therefore, the amorphous phase can be reached by cooling rapidly to below the glass transition temperature. In a phase-change optical disc, rapid cooling is made possible by the small volume that is amorphized and the special stacking of layers comprising the disc. This will be discussed in greater detail in chapter 3.

The glass, or amorphous solid, is a solid like any other if macroscopic properties like shape, shear stiffness etc. are considered. However, on an atomic scale there is a difference between crystalline and amorphous solids. This is illustrated in Figure 16. For the crystalline material, the atoms (or groups of atoms) are arranged in a pattern that repeats periodically in three dimensions. The amorphous material does not possess this long-range order (periodicity). On a local scale, though, there is a high degree of correlation, or short-range order, similar to crystalline solids. For example, all atoms in Figure 16(b) have three nearest neighbors at nearly the same distance. For covalent, tetrahedral semiconductors such as Si, Ge or GaAs the short-range order in the crystalline and the amorphous state is very similar. On the contrary, it has been discovered recently, that the well-known phase-change material $Ge_2Sb_2Te_5$ shows a substantial difference in short-range order between the amorphous and the crystalline state. [62], [65] It still needs to be clarified if this is a generic feature of phase change materials.

(a)   (b)

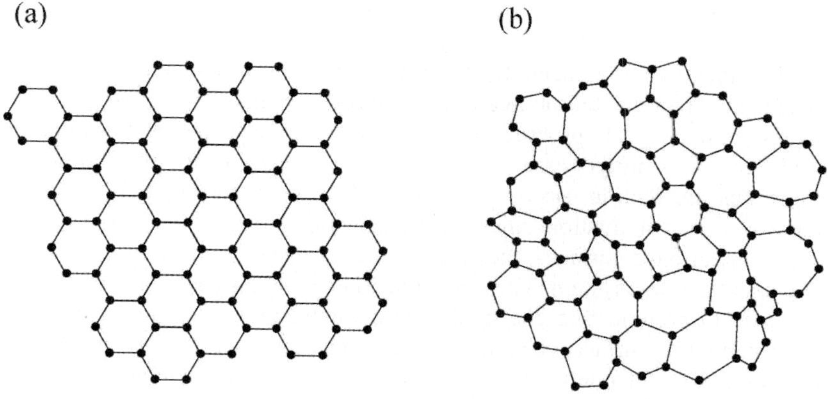

Figure 16. schematic representations of a crystalline and an amorphous structure.

Figure 17 shows a transmission electron microscope (TEM) picture of amorphous marks in a crystalline phase-change layer. In the crystalline phase, crystals with a fishbone-like diffraction contrast are observed. These variations in diffraction contrast are due to slight changes in crystallographic orientation. The amorphous area is smooth gray in Figure 17, no crystallographic contrast is observed (as there are no crystallites). Possibly, needless to say that Figure 17 shows the contrast between the amorphous and crystalline state for *electrons*. Optically this contrast is much more pronounced.

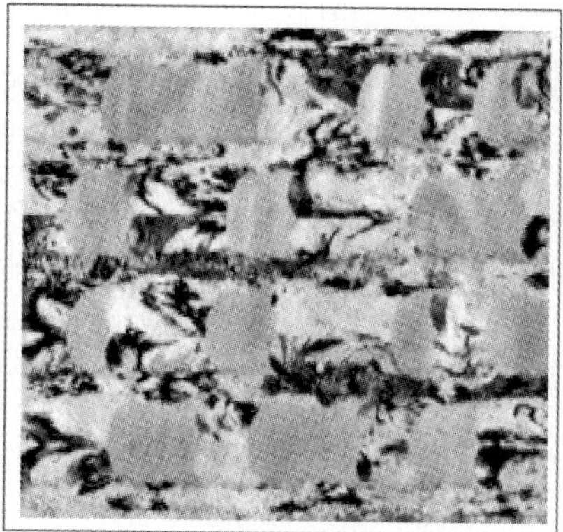

Figure 17. TEM picture of amorphous data in a crystalline phase-change layer.

2.3.   Crystallization

2.3.1.   Crystallization theory

Generally, crystallization occurs by a mechanism of nucleation and growth. In such a mechanism, small crystalline nuclei form initially, which subsequently grow. The formation of nuclei can proceed at the surface/interface and/or the bulk of the material. In the former case we speak of heterogeneous nucleation, while homogeneous nucleation occurs in the bulk. To understand the kinetics of crystallization, the activation energy for nucleation needs to be considered first. Since the nuclei have surfaces, the surface energy makes a positive contribution to the free energy of the system. The net change in free energy is therefore the sum of the decrease in volume free energy due to crystallization and the increase in free energy due to the surface energy, or better, the interface energy of the nuclei. Furthermore, a strain energy term may be present if the volume changes upon crystallization. The nucleus radius at which nuclei are kinetically stable, and growth is favored over dissolution of the nucleus, is called the critical size of the nucleus.

Experimentally, it is easiest to measure the overall rate of transformation instead of isolating nucleation and growth stages. Johnson, Mehl, and Avrami [22], [23], [24], [25], [26] found that the crystalline fraction χ could be described by

$$\chi(t) = 1\text{-}exp[-(k * t)^n], \qquad\qquad [1]$$

where n is the Avrami-exponent, a constant whose value depends on the nature of the nucleation and growth process, k is the rate constant and t is time. The rate constant k is described by

$$k = (1/3\ \pi u^3 * I)^{1/4}, \qquad\qquad [2]$$

where u is the growth rate and I the nucleation rate. To obtain this equation a 3-dimensional growth mode and a constant nucleation rate are assumed, leading to $n=4$. A qualitatively similar expression with an Avrami exponent $3<n<4$ is valid if the nucleation rate decreases with time. For temperatures below the melting point, the growth rate u and nucleation rate I can be expressed through their corresponding activation barriers E_I and attempt-frequencies ν_i.

$$u = \nu_u * exp(-E_{growth} / kT)$$
$$I = \nu_i * exp(-E_{nucleation} /kT) \qquad\qquad [3]$$

The activation barrier for growth can be identified as the activation barrier for self-diffusion E_a. This is the activation barrier depicted in Figure 15. Hence above T_G a modified version of the above equation should be used for most phase change materials, since they show fragile behavior, i.e. a viscosity as described by the Vogel-Fulcher law. Under the realistic assumption that crystallization is diffusion-limited the activation barrier for nucleation is the energy necessary for building a critical nucleus ΔG_c plus an additional diffusion step for adding one more atom.

$$E_{growth} = E_a$$
$$E_{nucleation} = \Delta G_c + E_a \tag{4}$$

Crystallization is controlled by the interplay of nucleation and growth. In this case a total activation barrier for crystallization E_{total} is observed.

$$k = v_k * exp(- E_{total} / kT) \tag{5}$$

Combining these equations yields an expression for E_{total}:

$$E_{total} = E_a + 1/4\ \Delta G_c \tag{6}$$

which is only valid under isothermal conditions. With this equation ΔG_c can be calculated when data for E_{total} and E_a are available for the temperature range of interest. The problem is that both the rate of diffusion and crystallization are so high at the temperatures of interest, *i.e.* for $T_g < T < T_m$ which describes the undercooled liquid state, that no successful measurements of those two quantities have been reported up to today.

Another possible way to determine the total activation barrier for crystallization E_{total} is by determining the crystallization temperature for different heating rates and by employing the Kissinger analysis. [27], [42], [50] For materials where crystallization proceeds via nucleation and growth, eq. (6) can be employed to determine the activation barrier for nucleation, once the activation barrier for growth, which also corresponds to the activation barrier for diffusion, is given. The Stokes-Einstein-equation

$$\eta * D \propto kT, \tag{7}$$

links the viscosity η and the coefficient of diffusion. Below T_g the material is frozen in an iso-configurational state. Hence, in a Newtonian viscous flow regime the temperature dependence of η is expressed by

$$\eta \propto kT * exp(Q_{iso} / kT), \tag{8}$$

where Q_{iso} is the iso-configurational activation energy. Such a behavior is typical for amorphous materials below the glass transition temperature, T_g, as can be seen in the schematic drawing of Figure 15. Above T_g, however, Vogel-Fulcher behavior characterizes phase-change materials. Comparing equations (7) and (8) with the temperature dependence of diffusion

$$D \propto exp(-E_a / kT),\qquad\qquad [9]$$

identifies

$$E_a = Q_{iso}\qquad\qquad [10]$$

and, thus, provides the means to measure E_a by determining the temperature dependence of the viscosity as shown in. [37] In Table 4, the activation energy for viscous flow in the amorphous state is listed for three different phase-change alloys, together with the melting temperature (liquidus temperature). A scaling relation is obtained, which implies that for the three alloys studied the viscosity is similar at the same relative temperature T/T_M. For T_g, it has been tacitly assumed that the crystallization temperature of the amorphous state is a good estimate of the glass transition temperature, since experimentally determined values for this temperature do not exist. It should be noted that the values given in table 4 depend considerably on the heating rate employed in the experiment.

Recently, crystal growth velocities and the activation barrier for growth have also been determined for temperatures slightly above T_g. [39], [40] These experiments reveal that although the three alloys in Table 4 show an activation barrier for crystal growth of similar magnitude, the scaling of the activation barrier with the melting temperature is less pronounced. The clear difference of the activation barriers above and below T_g should also be noticed.

Table 4. Activation energy for viscous flow above Tg (E_u) and iso-configurational viscosity (E_η) for three phase change alloys. [39], [40]

Alloy	E_u (eV)	E_η^a (eV)	T_M (K)	T_G (K)
AgInSbTe	2.90 ± 0.05	1.33 ± 0.09	810	430
$Ge_4Sb_1Te_5$	2.74 ± 0.03	1.94 ± 0.09	958	445
$Ge_2Sb_2Te_5$	2.35 ± 0.05	1.76 ± 0.05	903	405

2.3.2. *Crystallization of amorphous marks in a phase-change disc*

Crystallization of a completely amorphous phase-change film occurs through a process of nucleation and growth as described by the formulas given above. However, for (small) amorphous data marks in a crystalline phase-change layer the situation may be different. [49] When the probability for crystal growth is much larger than for nucleation, crystallization may occur by growth from the crystalline-amorphous mark edges. This is the case for some phase-change materials, as will be discussed in the next section. In this case, the crystallization rate can be simply expressed as

$$k = v_k * exp(- E_{growth} /kT) \qquad\qquad [11]$$

if the phase-change material forms a strong glass or T is below T_G.

Figure 18 shows schematically the probability for nucleation and crystal growth as a function of temperature for two types of phase-change materials. In Figure 18(a) crystallization of amorphous marks occurs through nucleation and growth, whereas in Figure 18(b) the probability for crystal growth is dominant over nucleation. It can be observed that below the glass transition temperature, T_g, no crystallization occurs due to the (too) low mobility of the atoms. Above the melting temperature, T_m, the fluid phase is more stable than the crystalline phase and there is no driving force for crystallization. Crystallization takes place between T_g and T_m. In this region, the atomic mobility increases with increasing temperature but the driving force for crystallization decreases, which explains the maximum in the crystallization probability.

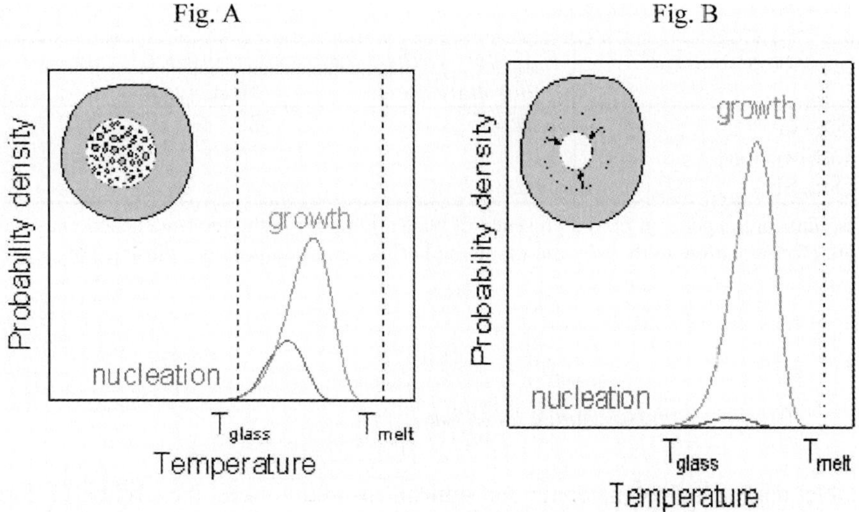

Figure 18. Probability for nucleation and growth as a function of temperature for amorphous marks in a crystalline phase-change layer. In Fig. A crystallization occurs by a process of nucleation and growth, in Fig. B crystallization proceeds by crystal growth from the amorphous-to-crystalline interface to the center of the amorphous region.

In rewritable optical recording, crystallization plays a dominant role. For example, old data are overwritten with new data after re-crystallization of the phase-change material with the laser. Here, the heat induced by the laser is so large that crystallization takes place at nanosecond timescales. Ways to determine the crystallization rate in this temperature regime are described in section 3.4. Sometimes, the amorphous marks re-crystallize slowly even at room temperature. If this happens, data read-out becomes more difficult and eventually the data will be lost. The resistance of amorphous marks against re-crystallization at room temperature or slightly elevated temperatures is described by the archival life stability. An indication of the archival life stability can be obtained by measuring the (isothermal) crystallization time of amorphous marks in a crystalline phase-change layer at various temperatures, followed by extrapolating these measurements to room temperature according to eq. (11). [47], [48] Table 5 shows typical values for the activation energy for crystallization and the archival life stability of various phase-change materials. Methods to determine the archival life stability are discussed in more detail in chapter 3. It should be noted that it is not possible to do the extrapolation the other way around, *i.e.* using values from measurements below T_g to estimate the crystallization rate at temperatures between T_g and T_m. The main reason for this is the drastic change in viscosity as was already observed in Fig 15. [58]

Table 5. Activation energy for crystallization of amorphous marks in a crystalline phase-change layer and calculated archival life stability at 50 °C.

Phase-change material	E_{total} *for written amorphous marks (eV)*	*Archival life stability at 50 °C (years)*
$Ge_2Sb_2Te_5$	1.64[*]	3[*]
$Ge_3In_4Sb_{73}Te_{20}$	3.65	$5*10^6$
$Ge_{13}Sn_{20}Sb_{67}$	4.02	$2*10^{10}$

[*]These data are dependent on the presence of nucleation sites in the melt-quenched phase and are therefore sensitive to the preparation method of the amorphous phase, presence of oxygen, *etc.*

2.4.   *Classes of phase-change materials*

Suitable phase-change materials for optical recording have the ability to form glasses at timescales and temperatures that are appropriate for given conditions such as data transfer rate and available laser power. This requires that the glass formation takes typically less than 100 ns within an adequately cooled recording stack, and that the melting point is below 1000 °C, typically around 600 °C. Furthermore, the materials should possess sufficient optical contrast between the amorphous and crystalline state. Some materials that fulfill these demands and that are currently applied in rewritable CDs and DVDs, are indicated in Figure 19. The main constituents of these materials are antimony, tellurium and germanium. The phase-change materials are divided into two classes with slightly different compositions, based on their crystallization mechanism. [49] Below, materials properties of both phase-change classes will be discussed.

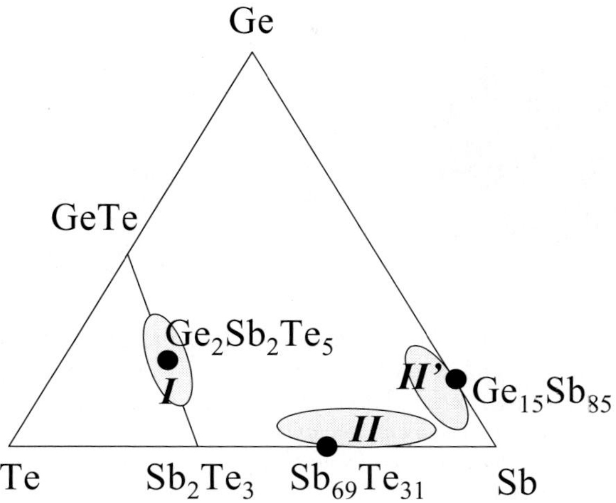

Figure 19. Composition triangle of Sb, Te and Ge. Compositions of materials with a nucleation-dominated crystallization mechanism, class I, and with a growth-dominated crystallization mechanism, class II, are indicated.

2.4.1. Class I: Crystallization by nucleation and growth

Materials of Class I are situated along the Sb$_2$Te$_3$–GeTe composition tie line in the ternary alloy phase diagram, and are often called stoichiometric materials. [50] The most important compound of this class is Ge$_2$Sb$_2$Te$_5$, whose trigonal crystal structure is depicted on the right-hand side in Fig. 20. It consists of a sequence of hexagonal layers of Ge, Sb and Te. However, if short laser pulses are used to crystallize this material, a metastable structure is formed, which is characterized by six fold coordination of the atoms with the cubic arrangement characteristic for the rocksalt lattice. This structural arrangement is shown on the left-hand side in Fig. 20. The Te atoms occupy each site of their fcc sub-lattice in the NaCl structure. Alternating Ge, Sb, and vacancies occupy the sites of the other sub-lattice. Recent studies [63], [65], [67] find compelling evidence for a pronounced local distortion away from the six fold coordinated sites, which considerably lowers the energy of the solid. [65] There is a high activation barrier for the rearrangement from the metastable rocksalt structure to the stable trigonal structure. [28], [29] The reason for this high barrier can be understood from Figure 21, which shows the rearrangement that is necessary to obtain the trigonal structure from the cubic structure. This massive rearrangement is not possible to realize during the short laser pulses that are typically employed in optical recording. Hence, in optical media based on $Ge_2Sb_2Te_5$ the metastable NaCl-structure is usually formed.

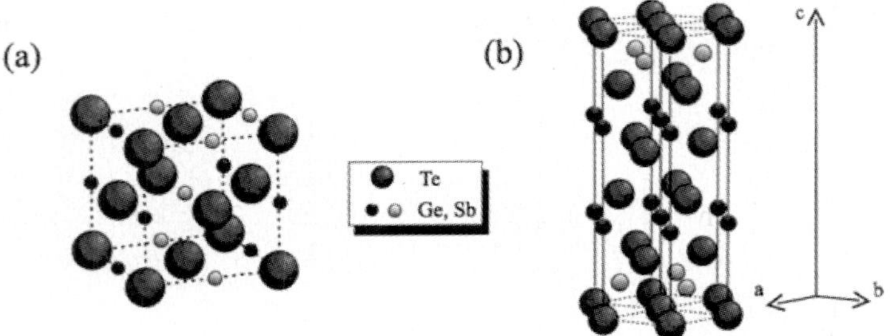

Figure 20. Two crystalline phases of $Ge_2Sb_2Te_5$. On the left hand side the metastable rocksalt structure is depicted. This phase is obtained by rapid crystallization of $Ge_2Sb_2Te_5$ such as upon annealing by short laser pulses. The Te atoms occupy each site of their sub-lattice in the NaCl structure. Ge, Sb, and vacancies occupy the sites of the other sub-lattice. The stable crystalline structure of $Ge_2Sb_2Te_5$ is shown on the right hand side. Hexagonal layers of Ge, Sb and Te characterize the structural order, where every species occupies one sub lattice in the a-b plane.

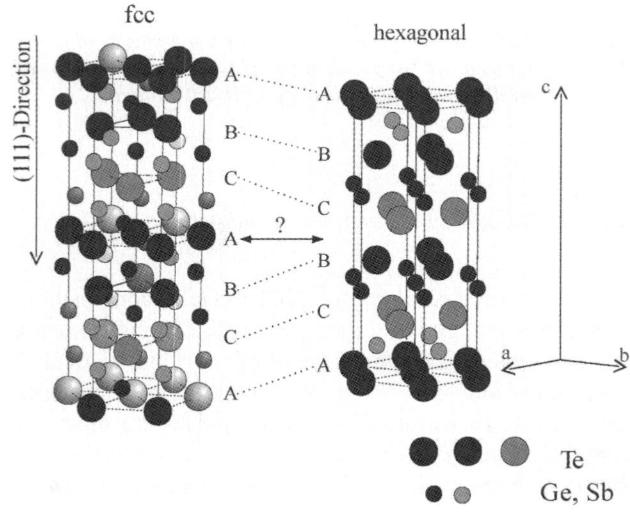

Figure 21. Mechanism of rearrangement for $Ge_2Sb_2Te_5$ from the metastable cubic (fcc) to the stable trigonal (hexagonal) phase. The rearrangement is accomplished by chemical ordering on the Ge, Sb sub- lattice and removal of a part of the stack. This massive rearrangement requires considerable atomic diffusion, which explains both the high activation barrier for this process and why it is impossible to reach the stable structure by irradiation with short laser pulses.

A schematic representation of the crystallization process of amorphous marks in a crystalline $Ge_2Sb_2Te_5$ layer is given in Figure 22(a). Crystallization is characterized by nucleation, followed by growth of the nuclei over a small distance, until they impinge upon other crystallites. Due to this crystallization mechanism, these materials are often referred to as nucleation-dominant materials (NDM). The high nucleation probability typical for this class of materials has implications for the morphology of the crystalline layer, as can be observed in Figure 22(c). A large number of small crystallites are observed. Also in the write process, which involves melt quenching of the phase-change material, some re-crystallization of the molten state may occur. This may have a significant effect on the shape of recorded marks. To prevent the formation of strangely shaped marks as a result of incomplete re-crystallization, influencing the nucleation probability of the phase-change material may be necessary.

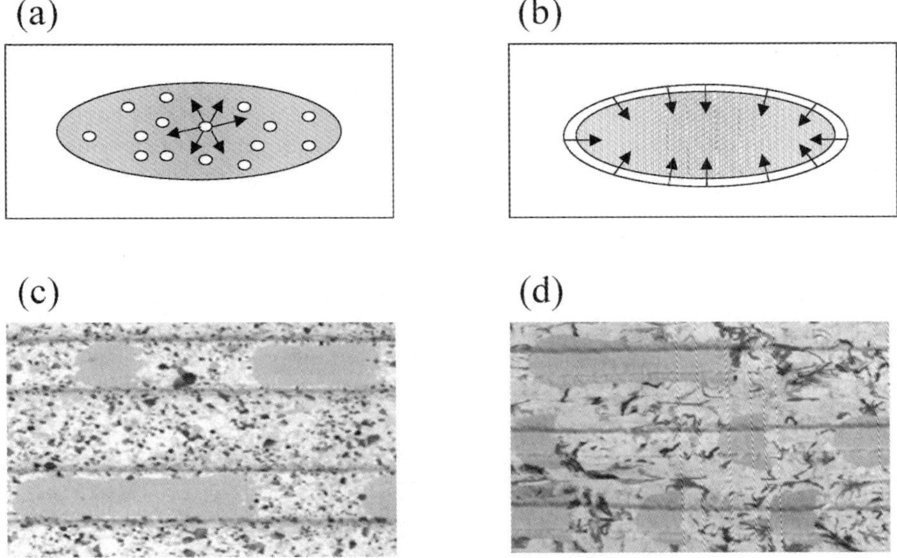

Figure 22. (a), (b) Schematic representation of the crystallization of an amorphous mark in a crystalline phase-change layer of a material with a nucleation-dominated crystallization mechanism (a) and with a growth-dominated crystallization mechanism (b). (c), (d) Transmission electron microscope images of amorphous marks in a crystalline phase-change layer. In (c) crystallization occurs via nucleation and growth, in (d) the crystallization process is dominated by crystal growth. The nature of the crystallization process can be clearly observed in the resulting crystalline texture.

The crystallization rate of class I phase-change materials can be increased by promotion of the nucleation probability. This can be done in several ways. [51] First, the crystallization rate increases with increasing phase-change layer thickness. This can be understood as a result of competition between the contributions of interface and bulk effects; the crystallization rate is dominated by the interface if the phase-change layer is thin and by the bulk if the layer becomes thicker than a critical value. Apparently, for $Ge_2Sb_2Te_5$ the nucleation probability in the bulk is higher than at the phase-change – dielectric interface of the recording stack for this range of layer thicknesses. Nucleation in the bulk can be promoted by addition of nitrogen or oxygen. Furthermore, adding nucleation promoting interface (or seed) layers, such as SiC, can also enhance the nucleation probability at the interface.

2.4.2.   Class II: Crystallization dominated by crystal growth

Figure 23 shows the phase diagram of the Sb-Te system. [52], [53], [54] On the horizontal axis the composition is given, ranging from 100% Te to 100% Sb. On the vertical axis the temperature is plotted. From such a diagram, possible compounds and solid solutions can be deduced at a certain temperature and composition. By substituting Sb for Te, the melting point is depressed and reaches a minimum at the composition $Sb_{69}Te_{31}$. This is the so-called eutectic composition. In general, compositions near the eutectic favor glass formation since the lower melting point causes the liquid to be less undercooled at the glass transition temperature, which reduces the possibility for crystallization. Furthermore, the lower melting temperature near the eutectic makes writing of amorphous marks in the material possible with commercial low power laser diodes.

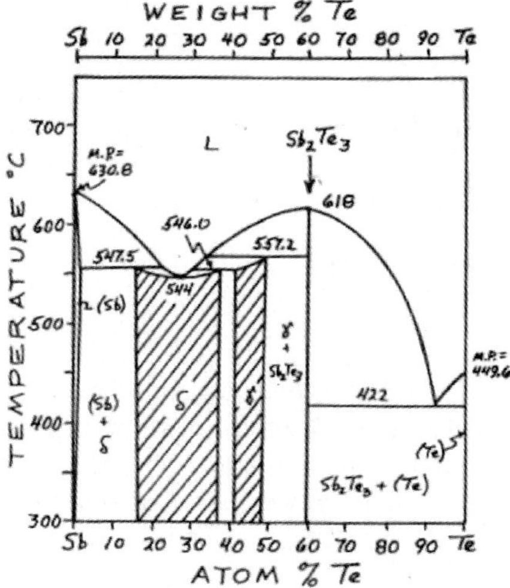

Figure 23. Phase diagram of the Sb-Te system. The horizontal axis shows the composition, from 100 % Sb on the left to 100 % Te on the right. [52], [53], [54]

Figure 24 shows the crystal structure of Sb_2Te. [55] Laser annealed films of Sb_2Te doped with Ag, In or Ge usually crystallize into a layered structure that is also found for As, Sb and Bi. It can well be visualized as derived from a cubic closed-packed structure. The main difference between the cubic closed packed structure and the structure in Figure 24 is that the layer distances between the different hexagonal planes differ from plane to plane in Figure 24.

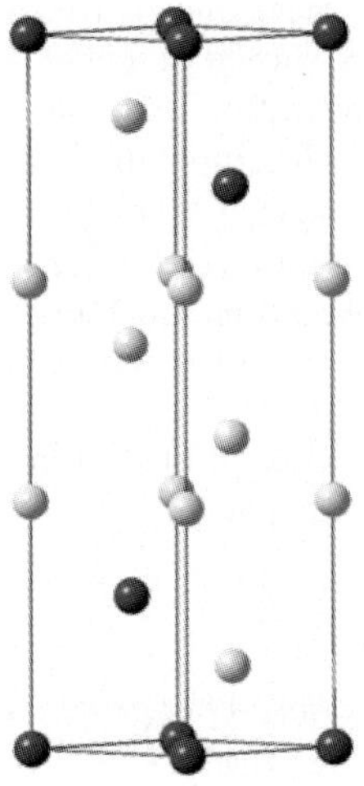

Figure 24. Crystal structure of Sb₂Te.

Figure 22(b) shows a schematic representation of the process of re-crystallizing amorphous marks in class II materials. [49] The crystallization mechanism is characterized by crystal growth from the amorphous-crystalline interface. Nucleation is very slow, on the order of μs, whereas crystal growth may occur on a ns timescale. Phase-change materials of class II are, therefore, also called growth-dominant materials (GDM) or fast growth materials (FGM). Figure 22(d) shows a transmission electron microscope (TEM) picture of a crystalline Ge-doped Sb-Te layer. Large and irregular shaped crystallites are observed, characteristic for the growth-dominated crystallization mechanism. FG materials have the advantage that for a given composition and, hence, a given crystallization rate, the time needed for re-crystallization of an amorphous mark decreases with the mark size. Thanks to this property discs based on FG materials have been shown to achieve high data rates (short crystallization time) in high data density formats (small amorphous marks), such as DVD and Blu-ray.

The properties of the class II materials can be adjusted by slight variations in the composition. [48] This is largely an empirical process. For example, increasing the Sb/Te ratio increases the crystallization rate, though, this occurs at the expense of the stability of the amorphous phase against crystallization (archival life stability). The material properties can be further modified by addition of dopant atoms. For instance, adding small amounts of Ge to the phase-change material enhances the amorphous phase stability.

One of the main aims in optical recording is to achieve high data transfer rates. To be able to do so, rewritable discs should be designed that incorporate phase-change materials having a high crystallization rate at elevated temperatures, but with virtually no crystallization at room temperature, in order to prevent the data from spontaneous erasure. For this purpose, a special (sub-) class of fast growth materials has proven to be promising (see region II' in Figure 19). This sub-class comprises of materials with compositions close to the Ge-Sb (or Ga-Sb) eutectic, where Sn may replace some of the Sb. [56], [57] Te or other atoms may be added as dopants. The materials properties of these compositions can be adjusted in a way similar to material optimization of Sb-Te compositions. More about recording media for high-speed data storage can be read in chapter 5.

2.5.   *Emerging directions in phase-change research*

Phase change materials have been developed over the course of the last thirty years by a combination of intuition, insight into material properties and their dependence upon stoichiometry as well as trial and error approaches. In the near future phase-change media are even facing more daunting tasks. The materials should be suitable for faster and faster data storage, yet still provide the required stability at room temperature. In addition, the phase-change materials should preferably also enable multilayer or multilevel recording.

One approach that allows fast development of suitable phase-change materials is combinatorial material synthesis. This approach has found widespread use in areas as diverse as the development of pharmaceutical substances and phosphors for lighting applications. For the development of phase-change materials several groups have tried concepts of combinatorial material synthesis. [60], [61], [64] An efficient way to rapidly identify the optimum composition of known phase-change components is co-sputtering. In this approach, a rotating magnetron confines the plasma to a sector of a segmented target. During rotation of the magnetron, the sputter power to the plasma is varied from segment to segment. By adjusting these power settings the composition of the sputtered layer can be varied over a relatively wide range. [66] From such experiments a detailed understanding of the correlation of stoichiometry with important material parameters such as optical contrast or the minimum time of re-crystallization can be obtained. However, it has to be stressed, that this approach needs to be combined with efficient schemes to determine materials properties and to identify general trends in the large databases created by combinatorial approaches.

At the same time, research is focusing on determination of the structure of the amorphous phase and understanding the relationship between atomic arrangements and optical contrast. [62] The optical contrast between the amorphous and the crystalline state is one of the mandatory requirements phase change media need to meet. Since the change of optical properties is caused by an atomic rearrangement, this rearrangement needs to be large enough to produce good contrast. This is schematically depicted in Figure 25. On the other hand, a large structural rearrangement can lead to long times needed for re-crystallization if long range diffusion is a prerequisite. It is, hence, not trivial to find materials that combine a high optical contrast and a fast atomic rearrangement.

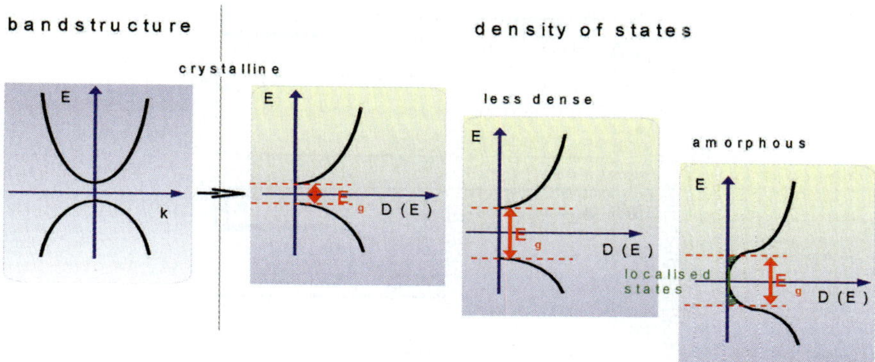

Figure 25. Schematic change of the density of states upon amorphization of a crystalline phase change material. For chalcogenide based phase change alloys a reduction of density as encountered in the amorphous state leads to an increase in band gap. The disorder characteristic for the amorphous state in addition leads to the appearance of localized states in the band gap.

On the left hand side a highly simplified 'band-structure' is depicted for a crystalline material. This leads to a density of states as shown on the next diagram to the right. Upon the transition to the amorphous state, the density of the atoms is reduced. The reduced overlap of the electronic wave functions leads to an opening of the gap for the phase-change materials. This opening of the gap should be related to the change in atomic volume or density, respectively. Finally amorphization is also linked with the creation of defect states below the edge of the conduction band and above the edge of the valence band leading to so-called tail states. Recent research activities focus on the question if these two changes are sufficient to explain the optical contrast between the amorphous and crystalline state in phase change media.

In fact, phase-change materials that are employed in optical storage do exhibit a pronounced density (volume) change upon phase transition. This is demonstrated in Figure 26 for three phase change alloys, where the relative film thickness is shown after annealing for 10 min to different temperatures. The films, which had a thickness between 35 and 60 nm were sputter deposited onto Si wafers. To facilitate the comparison of the data, the thickness is normalized with respect to the thickness of the as-deposited film. For the three different alloys studied here, crystallization occurs around 155°C for AgInSbTe, 130°C for $Ge_2Sb_2Te_5$ and 170°C for $Ge_4Sb_1Te_5$. In all cases crystallization is accompanied by a considerable reduction in film thickness. The smallest thickness decrease with 5.5% is observed for AgInSbTe, while $Ge_2Sb_2Te_5$ shows a 6.5% thickness decrease upon crystallization. $Ge_4Sb_1Te_5$ is even characterized by a 9.0% thickness reduction.

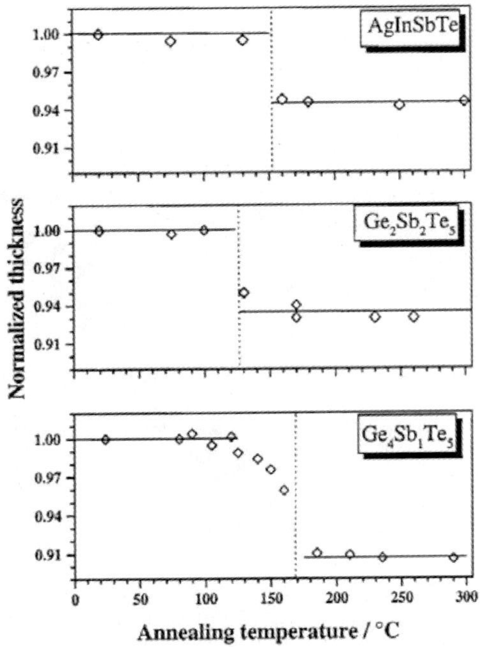

Figure 26. Film thickness of AgInSbTe, $Ge_2Sb_2Te_5$, and $Ge_4Sb_1Te_5$ films as a function of increasing annealing temperature as measured by x-ray reflectometry. [46] Crystallization, which leads to a sudden decrease in film thickness, is observed at 155°C for AgInSbTe, 130°C for $Ge_2Sb_2Te_5$, and 170°C for $Ge_4Sb_1Te_5$. The more gradual thickness change upon annealing for $Ge_4Sb_1Te_5$ is partly due to the formation of a thin oxide film. To facilitate a comparison of different data sets, all thicknesses are normalized with respect to the thickness of the as-deposited film. Crystallization leads to a 5.5 % thickness decrease for AgInSbTe, a 6.5 % thickness decrease for $Ge_2Sb_2Te_5$, and a 9 % thickness reduction for $Ge_4Sb_1Te_5$.

Hence, it seems reasonable to assume that sufficient optical contrast can only be obtained if the transformation from the crystalline to the amorphous state is accompanied by a considerable structural rearrangement, such as a pronounced density change. This is confirmed by the data compiled in Table 6, which contains a list of materials that show good optical contrast together with a number of other relevant parameters. For all phase-change materials with sufficient optical contrast, a considerable density change of 5% or more is observed. On the other hand, the materials with a low optical contrast show a rather low-density change upon crystallization of 1-2 %.

These data suggest that good contrast can only be found for those materials where phase transition is accompanied by a pronounced density change. Furthermore, all materials exhibiting a large density change, and hence a high optical contrast, have a cubic or cubic-like structure. The term cubic structures is used to describe an octahedral arrangement of atoms. In the cubic-like structures the building blocks are moderately distorted octahedral.

The fact that materials with a meta-stable or stable cubic or rocksalt structure are potentially promising candidates for phase-change recording has been pointed out, for a different reason, by Yamada and coworkers [45] who stated that cubic structures are preferable in high-speed phase-change materials since they enable particularly fast crystallization due to their simple crystalline structure. Peculiarly, the chalcopyrite or the related wurzite structure is also relatively simple. Nevertheless, as Table 6 reveals, they are not suitable for phase-change recording since they do not show a sufficient optical contrast. In this sense, it would be helpful to understand why cubic-like (octahedral) structures are prevalent in phase-change alloys while tetrahedrally coordinated alloys appear to be unsuitable for phase-change recording.

Table 6 shows experimental results, which demonstrate that for structures with an average number of valence electrons (N_{sp}) very close to 4, sp^3-bonded structures are stable, as observed for Si, Ge, GaAs and InP. The underlying cause for this structure is the sp^3-hybrid formed to maximize the overlap of the atomic orbitals. This hybrid is very favorable for compounds where every atom on average has 4 valence electrons. In this case the bonding fraction of the sp^3-hybrid is completely occupied while the antibonding sp^3-orbital is completely empty. For compounds with an average number of valence electrons larger than 4, the antibonding sp^3- orbital will be partially occupied destabilizing the ZnS or diamond lattice, respectively. Hence, for a larger average number of valence electrons other structures, like the octahedral structures, become favorable. In this case the behavior of the p-electrons alone controls the structural arrangement with minimum energy.

*Table 5. PC materials with sufficient (**bold**) and insufficient (italics) optical contrast, their density change upon crystallization as measured by XRR, their structures and their average number N_{sp} of valence electrons. Please note that both $Ge_1Sb_2Te_4$ and $Ge_2Sb_2Te_5$ each have one vacancy per unit cell in the cubic structure, but not in the hexagonal structure. This explains the different average number of valence electrons for $Ge_1Sb_2Te_4$ with cubic and hexagonal structure.*

Material	Structure	Optical constants at 650 nm		N_{sp}	$\Delta\rho$ %
		Crystalline n, k	Amorphous n, k		
GeTe	Rock salt	6.1, 0.2 [70]	3.6, 1.5 [70]	5.0	-
Ge₁Sb₂Te₄	Rock salt (metastable)	3.6, 4.2 [70]	4.1, 2.1 [70]	4.75	7.5 [70]
	Hexagonal (stable) [42]			5.4	8.2
Ge₂Sb₂Te₅	Rock salt [32]	4.4, 4.3 [68]	4.5, 2.3; 4.2, 1.9 [68], 4.1, 2.1 [69]	4.8	6.8 [35]
	Hexagonal [35]	3.9, 4.3 [69]			8.2
Ge₁Sb₁Te₅	Rock salt [36]	3.2, 3.9	3.8, 1.3	5.1	9.3
	Rock salt [43]			4.3	[36][36]
AgSbTe₂	Rock salt [38]	2.7, 3.5	3.2, 2.8	4.5	4.3
AuSbTe₂	Rock salt [44]	3.1, 4.2	3.3, 3.2	4.5	5.0
Au₂₅Ge₄Sn₁₁Te₆₀	Cubic [45]			4.45	-
Ag₃In₄Sb₇₆Te₁₇	Cubic [45]			4.93	-
Au₅₅Ge₆.₅Sn₅₉Te₂₉	Cubic-like/ hexagonal	2.4, 3.7	4.6, 3.0	4.94	5.2 [35]
AgInTe₂	Chalcopyrite [38]	3.3, 0.7	3.3, 0.6	4.0	1.7 [38]
AuInTe₂	Chalcopyrite [44]	3.3, 1.2	3.8, 1.1	4.0	2.0 [44]

This is nicely demonstrated by density functional theory. As an example, results of calculations for AuSbTe₂ (N_{sp} = 4.5) are given in Figure 27(a). The calculations reveal the p-bonding type. Accumulation of bond charge in the direction of 4 nearest neighbors can be seen in the (001) plane around the Te atom. The two remaining nearest neighbors reside in the planes above and below the reference plane. This six-fold coordination in a covalent material is the characteristic feature of p-bonding. For the (110) plane of AuInTe₂ (N_{sp} = 4.0) sp³-hybridization is observed. These findings are summarized in Figure 28, which displays the energy difference per single atom as a function of the average valence electron number for different alloys.

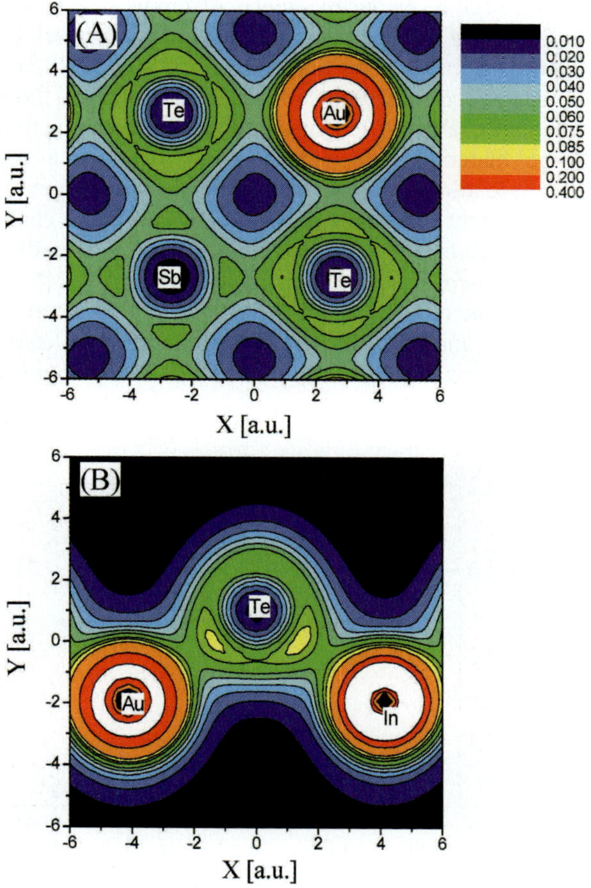

Figure 27. The electronic density in (A) for the (001) plane of AuSbTe$_2$ with rocksalt structure and in (B) for the (110) plane of AuInTe$_2$ with chalcopyrite structure. We consider the following electrons as valence electrons: 5d^{10}6s^1, 4d^{10}5s^{2}5p^1, 5s^{2}5p^3, 5s^{2}5p^4 for Au, In, Sb and Te, respectively to create the pseudopotential used in the DFT calculation. The electronic density is in unit of e/a.u.^3, where a.u. corresponds to Bohr's radius. [41]

For all calculated Te based ternary alloys, it is found that when $N_{sp} = 4$ the alloys prefer the chalcopyrite structure with a relatively large ΔE. Increasing N_{sp} will decrease ΔE. When $N_{sp} = 4.25$, the change of sign for the energy difference ΔE indicates that the alloy prefers the rocksalt structure. Even though this result does not imply that alloys like AgSnTe$_2$ and AuSnTe$_2$ ($N_{sp} = 4.25$) will have a rocksalt structure, since other crystalline phases might even have a lower energy, the chalcopyrite structure can be excluded. Hence, materials such as $CuSnTe_2$, $AgSnTe_2$ as well as $AuSnTe_2$ might also be suitable candidates for phase-change recording if we regard the optical contrast restriction only. From Figure 28, an estimate can be made revealing that the critical value of N_{sp} distinguishing the rocksalt and the chalcopyrite structures is about 4.1. This simple criterion can facilitate the search for new phase-change materials and may open the way to a more fundamental understanding of these alloys. Yet for such in-depth knowledge more emphasis has to be put on an insight into the structure and property change between the amorphous and crystalline state. Recent studies show compelling evidence of a pronounced change of *local* structure upon crystallization. [62], [63] This observation could help to provide the required input for calculations to determine the properties of both the amorphous and crystalline state with ab-initio calculations and, hence, open a new area of materials design for phase-change media applications.

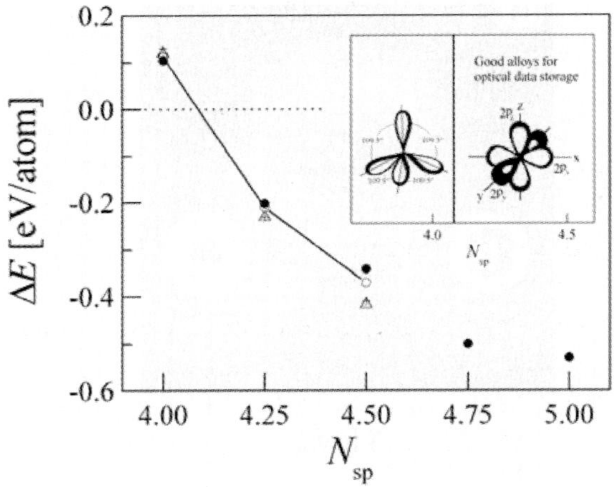

Figure 28. Plot of energy difference per single atom ΔE for different average number of valence electrons N$_{sp}$. Open triangles (Δ) represent Au(In,Sn,Sb)Te$_2$; Open circles (○) represent Ag(In,Sn,Sb)Te$_2$; Crosses (+) represent Cu(In,Sn,Sb)Te$_2$; Solid circles (●) represent AuGaTe2, AuGeTe2, AuAsTe2, InSnTe2, and InSbTe2, respectively. Solid line is the link for Ag series of alloys. The border separating the chalcopyrite and rocksalt structure is estimated as about N$_{sp}$ = 4.1. The insert shows the different bonding in materials crossing the border. [41]

2.6.   *References of chapter 2*

[16]   M. Libera and M. Chen, MRS Bull. 15, 40 (1990)

[17]   S. R. Ovshinsky, Phys. Rev. Letters 21, 20 (1968)

[18]   N. Yamada, MRS Bull. 1996, 21, 48

[19]   H. Vogel, Das Temperaturabhängigkeitsgesetz der Viskosität von Flüssigkeiten. Phys. Z., 22, 645 (1921)

[20]   G.S. Fulcher, J. Am. Ceram. Soc., &, 339 (1925)

[21]   F. Spaepen, Mat. Res. Soc. Symp. Proc. 57, 161 (1986)

[22]   W.A. Johnson and R.F. Mehl, Trans. Am. Crytallogr. Assoc. 135, 416 (1939)

[23]   M. Avrami, J. Phys. Chem. 7, 1103 (1939)

[24]   M. Avrami, J. Phys. Chem. 8, 212 (1940)

[25]   M. Avrami, J. Phys. Chem. 9, 177 (1941)

[26]   G. Ruitenberg, A.K. Petford-Long and R.C. Doole, J. Appl. Phys. 92, 3116 (2002)

[27]   H.E. Kissinger, Anal. Chem. 29, 1702 (1957)

[28]   V. Weidenhof, I: Friedrich, S. Ziegler und M. Wuttig, Atomic force microscopy study of laser induced phase transitions in $Ge_2Sb_2Te_5$, J. Applied Physics 86, 5879 (1999)

[29]   I. Friedrich, V. Weidenhof, W. Njoroge, P. Franz und M. Wuttig, Structural transformations of $Ge_2Sb_2Te_5$ films studied by electrical resistance measurements, J. Applied Physics 87, 4130 (2000)

[30]   V. Weidenhof, N. Pirch, I. Friedrich, S. Ziegler und M. Wuttig, Minimum time for laser induced amorphization of $Ge_2Sb_2Te_5$ films, J. Applied Physics 88, 657 (2000)

[31]   V. Weidenhof, I. Friedrich, S. Ziegler und M. Wuttig, Laser induced crystallization of amorphous $Ge_2Sb_2Te_5$ films, J. Applied Physics 89, 3168 (2001)

[32]   I. Friedrich, V. Weidenhof, St. Lenk und M. Wuttig, Morphology and structure of laser-modified $Ge_2Sb_2Te_5$ films studied by transmission electron microscopy, Thin Solid Films 389, 239 (2001)

[33]   T. Pedersen, J. Kalb, W. Njoroge, D. Wamwangi, M. Wuttig, F. Spaepen, Mechanical stresses upon crystallization in phase change materials Applied Physics Letters 79, 3597 (2001)

[34]   W. Njoroge, M. Wuttig: Crystallization kinetics of sputter-deposited amorphous
        AgInSbTe films, J. Applied Phys. 90, 3816 (2001)

[35]   W. Njoroge, H.-W. Wöltgens, M. Wuttig, Density changes upon crystallization of
        $Ge_2Sb_{2.04}Te_{4.74}$ films, J. Vac. Sci. Technol. A 20, 230 (2002)

[36]   D. Wamwangi, W. Njoroge, M. Wuttig, Crystallization kinetics of $Ge_4Sb_1Te_5$ films
        Thin Solid Films 408, 310 (2002)

[37]   J.A. Kalb, F. Spaepen, M. Wuttig, Calorimetric measurements of phase transformations
        in thin films of amorphous Te alloys used for optical data storage, J. Applied
        Physics 93, 2389 (2003)

[38]   R. Detemple, D. Wamwangi, M. Wuttig, G. Bihlmayer, Identification of Te alloys with
        suitable phase change characteristics, Applied Physics Letters 83, 2572 (2003)

[39]   J.A. Kalb, F. Spaepen, T. Pedersen, M. Wuttig, Viscosity and elastic constants of
        amorphous Te alloys used for optical data storage, J. Applied Physics 94, 4908
        (2003)

[40]   J.A. Kalb, F. Spaepen, M. Wuttig, Applied Physics Letters 84, 5240 (2004)

[41]   Mengbo Luo and M. Wuttig, Dependence of crystal structure of Te based phase
        change materials on number of valence electrons, Advanced Materials 16, 439
        (2004)

[42]   Z. L. Mao, H. Chen, A. L. Jung, J. App. Phys. 78, 2338 (1995)

[43]   Y. Maeda, H. Andoh, I. Ikuta, H. Minemura, J. App. Phys. 64, 1715 (1988)

[44]   D. Wamwangi, X. Zhang, R. Detemple, H. Wöltgens, M. Wuttig, J. Appl. Phys. 64,
        1715 (2004)

[45]   T. Matsunaga, N. Yamada, Jpn. J. Appl. Phys. 41, 1674 (2002)

[46]   T. P. L. Pedersen, J. Kalb, W. K. Njoroge, D. Wamwangi, M. Wuttig, F. Spaepen,
        Appl. Phys. Lett. 79, 3597 (2001)

[47]   M. H. R. Lankhorst and H. J. Borg, Mat. Res. Soc. Symp. Proc. **674** , V1.4.1, (2001)

[48]   M. H. R. Lankhorst, L. van Pieterson, M. van Schijndel, B. A. J. Jacobs and J. C. N. Rijpers, Jpn. J. Appl. Phys. **42,** 863, (2003)

[49]   G. F. Zhou, H. J. Borg, J. C. N. Rijpers, M. H. R. Lankhorst and J. J. L. Horikx, Proc. of SPIE **4090,** 108, (2000)

[50]   N. Yamada, E. Ohno, K. Nishiuchi, N. Akahira and M. Takao, J. Appl. Phys. **69,** 2849 (1991)

[51]   G. F. Zhou, Mat. Sci. Eng. A, **A304-306,** 73, (2001)

[52]   N. K. Abrikosov, L. V. Poretskaya, I. P. Ivanova, Russ. J. Inorg. Chem. **4,** 1163, (1959)

[53]   S. Bordas, M. T. Clavaguera-Mora, B. Legendre, C. Hanchung, Thermochimica Acta **107,** 239, (1986)

[54]   G. Ghosh, H. L. Lukas, L. Delaey, Z. Mkde **80,** 731 (1989)

[55]   V. Agafonov, N. Rodier, R. Ceolin, R. Bellissent, C. Bergman, J. P. Gaspard, Acta Cryst. A **47,** 1141 (1991)

[56]   L. van Pieterson, M. van Schijndel, J. C. N. Rijpers and M. Kaiser, Appl. Phys. Lett. **83** 1373 (2003)

[57]   L. van Pieterson, J. C. N. Rijpers and J. Hellmig, Jpn. J. Appl. Phys. 43, 4974 (2004)

[58]   P.G. Debenedetti, F.H. Stillinger, Nature 410, 259 (2001)

[59]   C.A. Angell, K.L. Ngai, G.B. Mckenna, P.F. McMillan, S.W. Martin, J. Appl. Phys. 88, 3113 (2000)

[60]   S. Kyrsta, R. Cremer, D. Neuschütz, M. Laurenzis, P.H. Bolivar, H. Kurz, Thin Solid Films 398, 379 (2001)

[61]   R. Cremer, S. Dondorf, M. Hauck, D. Horbach, M. Kaiser, S. Kyrsta, O. Kyrylov, E. Münstermann, M. Philipps, K. Reichert, G. Strauch, Zeitschrift für Metallkunde 92, 1120 (2001)

[62]   A.V. Kolobov, P. Fons, J. Tominaga, A.L. Ankudinov, S.N. Yannopoulos, K.S. Andrikopoulos, Journal of Physics C 16, S5103 (2004)

[63]   A.V. Kolobov, P. Fons, A.I. Frenkel, A.L. Ankudinov, J. Tominaga, T. Uruga, NATURE MATERIALS 3 (10): 703-708 OCT 2004

[64]   M. Wuttig, R. Detemple, I. Friedrich, W. Njoroge, I. Thomas, V. Weidenhof, H.W. Wöltgens, S. Ziegler, Journal of Magn. Mag. Mat. 249, 492 (2002)

[65]   W. Welnic, A. Pamungkas, R. Detemple, C. Steimer, S. Blügel, M. Wuttig, NATURE MATERIALS 5, 56 (2006)

[66]   H.J. Borg, M.van Schijndel, J.C.N.Rijpers and M.H.R.Lankhorst, Jpn.J.Appl.Phys. 40 (2001) 1592-1597

[67]   S. Shamoto, N. Yamada, T. Matsunaga, T. Proffen, J.W: Richardson, J.H. Chung, J.H. Egami, Applied Physics Letters 86, 81904 (2005)

[68]   N. Yamada, M. Otoba, K. Kawahara, N. Miyagawa, H. Ohta, N. Akahira and T. Matsunaga Jpn. J. Appl. Phys 37, 2104-(1998)

[69]   S. Y. Kim, S. J. Kim, H. Seo and M.R. Kim, Jpn. J. Appl. Phys 37, 1713 (1999)

[70]   Han Willem Wöltgens, PhD Thesis (2003), RWTH Aachen

3.   Thermal modelling of phase-change recording

A detailed understanding of the mark formation and erasure mechanisms encountered in phase-change recording is essential to develop recording stacks and write strategies (see chapter 4). A multi-layer mark formation and erasure model was developed that has proven to be a predictive approximation of a pre-grooved phase-change disc (3.1). The optical properties of the films used in phase-change recording stacks, which are required input parameters for the thermal model, are discussed in section 3.2. Methods developed to determine some of the essential thermal and crystallization parameters for the simulation tool are described in 3.3 and 3.4, respectively. Typical modeling results obtained with the developed model are given in 3.5. The effect of the groove shape on the temperature distributions was investigated with a vector-diffraction model; the results are given in 3.6.

3.1.   *Multi-layer thermal model*

3.1.1.   *Heat diffusion in thin layers*

In metals, heat conduction is mainly dominated by electron transport, carrying heat from warmer to colder places. In dielectrics and most of the semi-conducting materials, heat conduction is phonon-dominated, induced by atom or ion vibrations in the lattice.

Due to the process of sputter-deposition, the thin sputtered dielectric layers are typically amorphous or microcrystalline. Phonon scattering at lattice imperfections is not the only cause of significant reduction in the thermal conductivity. The large differences with respect to the bulk material are also caused by the structural disorder (lattice defects) at the interfaces where the layer has started to grow from the sputter-deposition. This barrier, which may extend only a few angstrom within the film, poses a significant barrier to phonon transport. [74], [75], [76] Sputtered metal films typically have a columnar structure. [76], [77] The structural disorder also exists at metal interfaces.

The film thickness may have an important effect on the thermal conductivity. It is known that the film structure is a function of thickness. [76] The influence of the barrier at the interface becomes less important for thicker films. Measurements of the electrical conductivity of thin sputter-deposited films can also be used to illustrate the thickness dependence of the thermal conductivity. In-plane measurements of the in-plane electrical conductivity (or electrical resistivity) for two types of aluminum films are given in Figure 29 (made in two different sputter-deposition systems). Two important observations can be made. A strong dependence in electrical and thermal resistance exists, in particular for the films below 50 nm thickness. Furthermore, the electrical resistance of relatively thick films (1000 nm) does not approach that of bulk aluminum at all, indicating that the sputter-deposited film has a significantly different microstructure (the thermal conductivity of bulk aluminum is about 240 W/mK).

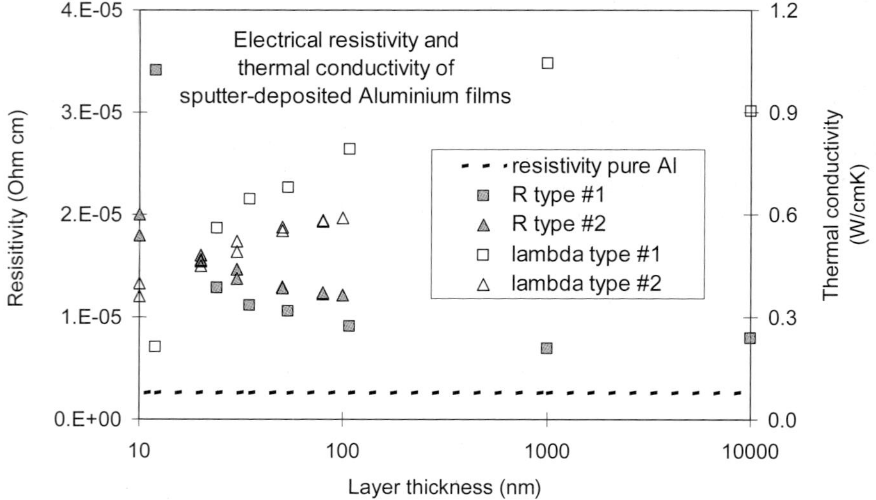

Figure 29. In-plane measurements of the electrical resistivity (R) and the derived thermal conductivity 'lambda' (via the Wiedemann-Franz law) as a function of the layer thickness for 2 types of sputter-deposited aluminum films. The dashed line indicates the resistivity of bulk Aluminum.

In summary, the two leading mechanisms for a reduced and possibly anisotropic thermal conductivity of thin films are 1) the structural disorder in the lattice and at interface and 2) the phonon scattering at the interface between the film and substrate.

3.1.2. Heat conduction at microscale

The conventional thermal diffusion equation is based on the assumption of infinite speed of heat propagation. A local change in temperature results instantaneously in a perturbation elsewhere in the medium, independent of the size of the medium. Fourier's law is in this concept a macroscopic description. If the time-scales are much larger than the typical relaxation times of the thermal waves, this assumption is justified. However, the physical time scales encountered in thin layer heat conduction may approach the relaxation times associated with phonon and electron collisions. To account for these finite propagation speeds, Cattaneo [79] and Vernotte [80] proposed the next modification of Fourier's law:

$$q + \tau \, \partial q / \partial t = - \lambda \, \nabla T,$$

where q is the heat flux, τ is the relaxation time, λ is the thermal conductivity and T is the temperature. Conservation of energy leads to the next hyperbolic diffusion equation: [81]

$$\tau \, \partial^2 T / \partial t^2 + \partial T / \partial t = D \, \nabla^2 T,$$

where D is the thermal diffusivity ($= \lambda / \rho c_p$, ρ is the density, c_p is the specific heat). If the term ($\tau \, \partial^2 T / \partial t^2$) is negligible, the conventional diffusion equation is obtained. Non-Fourier conduction with a possible source term (for example laser absorption) is accounted for with an additional time-dependent source term, multiplied with the relaxation time. [82], [83]

Based on the thermal relaxation times reported in the literature for the materials of interest, it is not expected that non-Fourier effects play an important role in the recording stacks considered for the current generation optical discs.

3.1.3.   Modelling of heat diffusion in multi-layer recording stacks

The temperature distribution in a multilayered stack due to laser heating by a moving laser spot can be described by the following heat diffusion equation, where the index n denotes the different layers: [85]

$$(\rho c_p)_n \, \partial T / \partial t = \lambda_n \, D \, \nabla^2 T + S_n(x,y,z,t),$$

ρ is the density, c_p is the specific heat (the product ρc_p is the heat capacity), λ is the thermal conductivity, T is the temperature increase and $S(x,y,z,t)$ is the heat source. Conservation of heat fluxes at the interface between two layers is a necessary boundary condition to couple the individual layers and to solve the heat diffusion problem.

The laser light absorption in a phase-change recording stack can be modeled in the following ways:

1) Scalar approach. The optical flux distribution in the propagation direction of the laser beam is based on a plane wave solution of the Maxwell equations in a multi-layer recording stack. The radial absorption distribution can be modeled as a Gaussian. [86], [87], [88] The optical properties of the different layers in the recording stack are needed to calculate the local absorption.

2) Rigorous vector diffraction. The dissipated laser power is calculated from vector diffraction in the grooved disc. [89], [90] This approach uses the finite

difference in the time domain (FDTD) method. First a source plane was defined with equivalent electric and magnetic sources. The E-field at the focal plane was calculated from Fourier transformation of the incident beam at the pupil entrance. The E-field at the source plane was obtained by forward propagation of light based on direct Fresnel diffraction theory. The incident magnetic field at the source plane was computed from Maxwell equations. The incident light at the source plane was used to calculate light diffraction in the grooved stack. The total absorbed laser light was calculated from the E-fields. This absorbed laser light is used as heat source in the heat diffusion equation. The indices of refraction and the absorption coefficients are required for each layer.

Direct heating of the layered structure can be modeled by the source term $S_n(x,y,z,t)$ that represents a moving heat source. The fraction of the incident light absorbed in the different layers is determined by the index of refraction (n) and absorption coefficient (k) of the layers in the stack. The shape of the heat source is determined by the optical system properties, such as the numerical aperture of the objective lens and the laser wavelength. In the model described here, a planar geometry is utilized and an assumption is made that the energy distribution in the light propagation direction can be calculated from the Poynting vector. [87], [88] It has been confirmed for planar geometries that the difference between the optical absorption calculated with such a scalar method and with a vector diffraction method is negligibly small. This becomes clear from the radial temperature profiles calculated for BD conditions (NA=0,85, 405 nm) and DVD conditions (NA=0.65, 658 nm) given in Figure 30. The scalar approach was implemented in the mark formation and erasure model, discussed in 3.1.4. The vector diffraction method was used to investigate the influence of groove geometry on the temperature distribution in the recording stack (3.6).

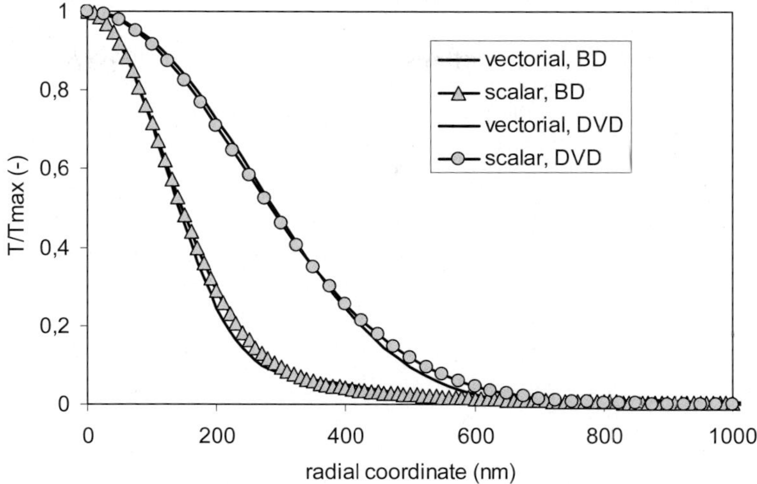

Figure 30. Radial temperature profiles calculated for a planar BD recording stack (NA=0.85, 405 nm) and DVD recording stack (NA=0.65, 658nm): comparison between scalar model and vector diffraction model.

Heat diffusion is described by the thermal properties of the different layers in the recording stack and substrate. For the sake of simplicity, it can be assumed that both the heat capacity and thermal conductivity of all materials are temperature-independent and that the properties are isotropic. Furthermore, the source term can be considered temperature-independent.

It is known that scattering of heat carriers, i.e. phonons and electrons, at lattice imperfections and interfaces causes significant deviations of the thermal conductivity while the heat capacity (ρc_p) is hardly influenced. The scattering of heat carriers in the layer itself can be seen as an intrinsic reduction of the thermal conductivity of the layer. The scattering at the layer interfaces is more a thermal interface resistance. The total thermal resistance of a thin layer, expressed as the effective thermal conductivity, then consists of the intrinsic thermal resistance and the additional interface effects. It should be explicitly noted here that separation of the two contributions to the effective thermal conductivity is not pursued since mainly descriptive thermal properties are required that provide an accurate prediction of the temperature distribution in a multi-layer recording stack.

The sputtering process and conditions determine to a great extent the interface properties. A more elaborate analysis is required if a separation of the intrinsic and interface contributions is pursued, for example by testing different sputtering conditions and materials as a function of the layer thickness.

3.1.4.   Mark-formation and erasure model

As indicated in chapter 2, basically two types of phase-change materials exist, namely the nucleation-controlled and the growth-controlled phase-change compositions. In this book, the focus is on fast-growth phase-change materials since these have been applied in all rewritable optical formats developed by Philips.

Several mark formation and re-crystallization models have been published to predict mark sizes during writing and erasing of data. [110], [111], [112], [107], [86] In the model described in [86] continuous crystal growth that occurs in the direct vicinity of the crystalline-amorphous interface is assumed. Atoms crossing the interface add themselves to the crystalline region, and vice versa. The frequency with which an individual atom is transferred from the amorphous to the crystalline phase is given by: [113]

$$\nu = \nu_0 \, e^{-Ea/R(T-Tg)},$$

where ν_0 is a characteristic frequency, E_a (J/mol) is the activation energy of transition from the amorphous to the crystalline phase and R is the gas constant (=8.314 J/mol/K). $T\text{-}T_g$ is the temperature difference between the interface temperature and the glass transition temperature. The glass transition temperature is introduced to obtain Vogel-Fulcher-Tammann behavior rather than ordinary Arrhenius behavior. [114] The frequency of the reverse transition is similarly:

$$\nu = \nu_0 \, e^{-Ea/R(T-Tg) - \Delta SM(TM-T)/RT},$$

where ΔS_M is the melt entropy (J/mol/K) and T_M is the melt temperature. The second term represents the free energy difference between the amorphous and the crystalline state. If the net transfer frequency is multiplied with a characteristic jump distance, the velocity of crystal growth can be obtained:

$$V(T) = V_0 \, e^{-Ea/R(T-Tg)} \, [\, 1 - e^{-\Delta SM(TM-T)/RT} \,],$$

with V_0 a pre-factor in μm/ns. Typical growth velocity curves are shown in Figure 31 for three values of V_0 and E_a. The other parameters in the growth velocity curve are T_M=833 K, T_g=425 K, and ΔS_M=22 J/mol/K. While V_0 determines the absolute value of the growth velocity, the width of the growth curve can be changed by the activation energy. The growth velocity drops to zero towards the glass transition temperature and is zero above the melting temperature.

The temperature-dependent velocity of crystal growth was implemented in the multi-layer thermal model (discussed in 3.1.3) to calculate the amount of re-crystallization for each time step. Mark formation was modeled as a thermal threshold phenomenon, namely amorphous material is formed when the melt temperature is exceeded. Let us consider the numerical procedure for mark formation and erasure in more detail. First, the thermal response to a write or erase laser power strategy is numerically calculated for the multi-layer recording stack. The time-dependent three-dimensional temperature distribution is then used to calculate the melt-edge in the phase-change layer, i.e. the area with a temperature, which exceeds the melting temperature of the phase-change material. The amount of re-crystallization, the growth distance, is also calculated from the temperature-dependent velocity of crystal growth *V(T)*. This crystal growth re-defines the amorphous-crystalline interface and leads to shrinkage of the amorphous mark. The total mark formation and subsequent crystal growth is calculated during the entire exposure to laser heating to ensure fully converged mark formation and erasure predictions.

A first estimate of the order of magnitude of the velocity of crystal growth can be derived from the erase specification. For BD-conditions, marks need to be erased at a recording velocity of 10 m/s (this is the 2X erase requirement). If erasure is carried out at this speed, a spot with 300 nm diameter (the 1/e-radius is 150 nm) corresponds to a dwell time of 300/10=30 ns. If 200 nm wide marks are written, 100 nm needs to be erased during passage of the laser spot (only half the mark width is used in the calculation since re-crystallization occurs from both sides). This corresponds to an average growth velocity of 100nm/30ns=3.3m/s. A similar estimate is obtained for DVD discs. The spot size is then 650 nm (the 1/e-radius is 320nm), which corresponds to a dwell time of about 640nm/8m/s=80ns at an erase velocity of 8 m/s. If 400 nm wide marks need to be erased, an average crystallization speed of 200nm/80ns=2.5 m/s is required.

The methods to determine the parameters that define the temperature-dependent velocity of crystal growth are discussed in section 3.4. One of the proposed methods utilize time-resolved static tester measurements in combination with numerical modeling. [115] In such experiments, marks of a pre-defined size are written and subsequently erased. A mark is formed in the crystalline recording layer if the temperature of the layer exceeds its melting temperature. Re-crystallization from the amorphous mark edge may occur whenever the temperature exceeds the threshold temperature of crystal growth but remains below the melting temperature, according to the crystal growth velocity curves given in Figure 31. In the case of growth-dominant material erasure of written amorphous marks can be handled as crystal growth from the amorphous mark edge. In that case, the temperature-dependent growth velocity can be determined from the crystallization time and the initial radius of the amorphous marks.

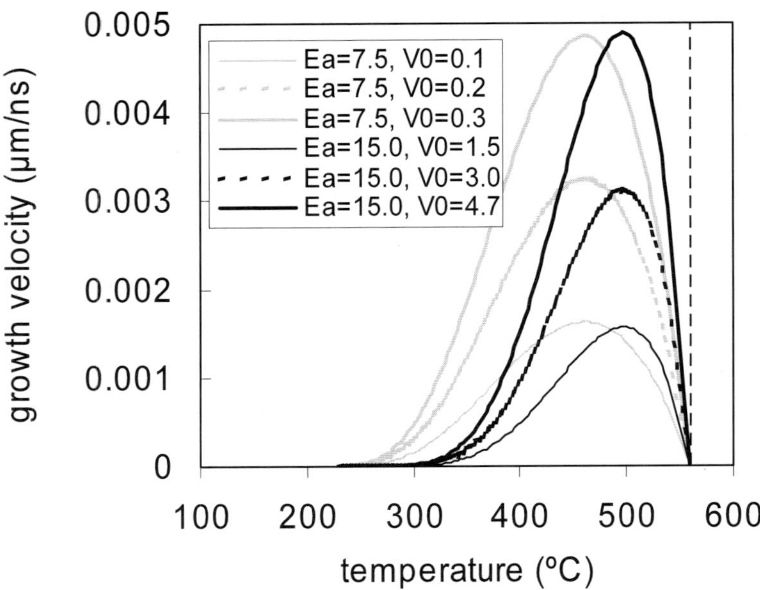

Figure 31. Typical growth velocity curves used in the crystallization model: curves for three values of the pre-factor V_0 (in µm/ns) and of the activation energy E_a (in kJ/mol).

3.2.   Optical parameters of phase-change recording films

The optical design of the stack is of decisive importance for both writing and readout of amorphous marks. The effective absorption of the incident laser power and, therefore, the attained write and erase temperatures, are to a large extent determined by the optical properties of the different layers in the recording stack. Furthermore, the quality of readout of the recorded data is determined by the contrast between the amorphous marks and the crystalline matrix, and by the absolute reflection levels of the two phases involved. Optimum performance is obtained by finding a compromise between optimum contrast, modulation and reflection and sufficient absorption to allow for writing and erasing of data.

The films used in optical discs are typically made with sputter-deposition techniques. The process conditions, such as the vacuum pressure, ambient gas, sputter power, substrate temperature etc, determine to a great extent the resulting layer morphology. For most of the sputter-deposited materials used in phase-change recording stacks, the resulting thin films differ from the corresponding bulk material. For example, sputter-deposited aluminum films grow more or less in column-like structures, gradually converging from separated columns to a polycrystalline structure. The multiple grain boundaries in the layer may lead to light refraction and scattering, thereby providing absorption and reflection characteristics that are different from the corresponding bulk material.

The index of refraction of a material is commonly expressed as a complex number $n=n\text{-}i\times k$, where n is the index of refraction and k is the extinction coefficient. The absorption of the material α is related to the extinction coefficient via $\alpha = 4\,\pi\,k/\lambda$ where λ is the wavelength of the laser light. Several established methods exist to measure the optical properties of thin films. Photometry and ellipsometry are among the most familiar methods. [71] Both techniques are based on reflection and transmission characteristics of the optical films under testing. It is beyond the scope of this book to provide a detailed description of the techniques. In the following paragraph, measurement results of some typical recording stack materials are given. These values of n and k determine the total laser power absorbed in the recording stack and were input parameters for the thermal simulation model.

3.2.1. *Optical properties of thin films*

Examples of photometric measurements are given for the materials that are most commonly used in phase-change recording stacks. [72] Typically, the wavelength-dependent reflection and transmission of a film are determined in a spectrophotometer. The measurements are fitted to a model to arrive at the n and k of the sample in question. For insulating materials, typically the Lorentz or empiric Cauchy model are used. Hagen Rubens model or the Drude model are more suitable for metal films.

The sputter-deposition process often influences the lattice structure, and therefore the optical properties of a thin film. The thickness dependence of n and k of silver are given in Figure 32. Results are plotted for 670 nm (DVD optics) and 405 nm (BD optics). For relatively thick silver films, above 30 nm thickness, the measured n and k values agree reasonably well with the corresponding bulk values. However, a significant deviation is observed for silver layers thinner than 30 nm.

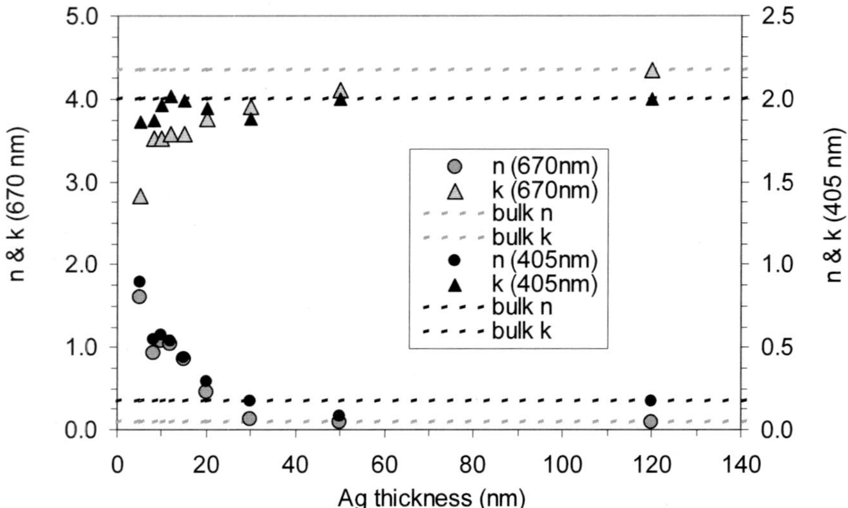

Figure 32. Index of refraction (n) and extinction coefficient (k) of sputter-deposited Ag as a function of the layer thickness for 405 and 670 nm wavelength.

ZnS-SiO_2 is most commonly used as dielectric layer. Measurements of *n* and *k* as a function of wavelength are given in Figure 33. The extinction coefficient is zero in the high-wavelength visible range, but becomes significant in the UV range, between 400 and 300 nm wavelength. For blue recording stacks, for example as used in Blu-ray Disc, the extinction coefficient is 0.002, which cannot be ignored in the optical design although it is still low. The steep increase at lower wavelength illustrates that it is quite difficult to use ZnS-SiO_2 at 257 or 266 nm wavelength. The index of refraction is rather high, gradually decreasing from 2.3 at 405 nm wavelength to 2.15 in the red wavelength range. These very-well matching optical properties in combination with the low thermal conductivity and the stability against high temperatures makes this material very suitable for phase-change recording stacks.

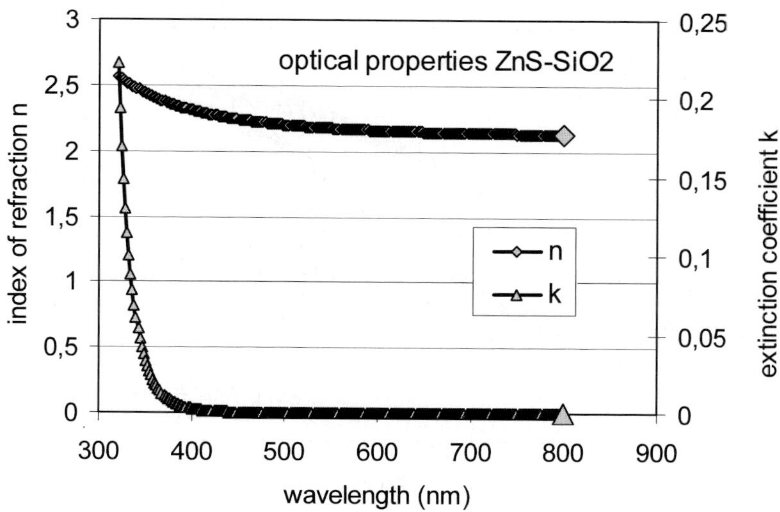

Figure 33. Index of refraction (n) and extinction coefficient (k) of ZnS-SiO$_2$ in the visible and near UV wavelength range.

Aluminum or silver are commonly used as metallic heat sink layers. Measurement results acquired for deposited films and literature values of *n* and *k* for bulk aluminum are compared in Figure 34. The measurements resemble the literature values quite well; the deviation can be explained from morphology difference due to the sputter-deposition process. Aluminum has a rather high extinction coefficient of 8 in the deep red region, which comes down to 5 at 400 nm wavelength. The index of refraction drops from 2.5 at 700 nm to 1.5 at 400 nm wavelength. Silver has a somewhat different behavior. Its extinction coefficient is lower than that of aluminum, and the index of refraction has a minimum around 450 nm wavelength.

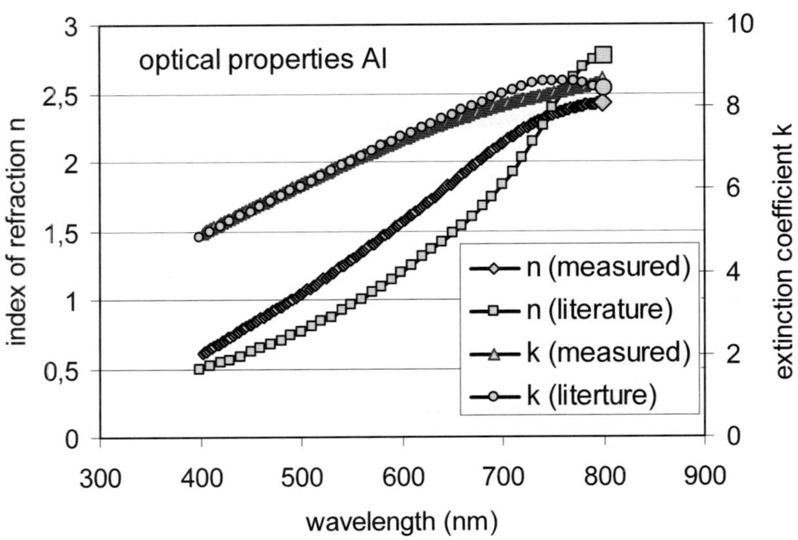

Figure 34. Index of refraction and extinction coefficient of aluminum as a function of the wavelength.

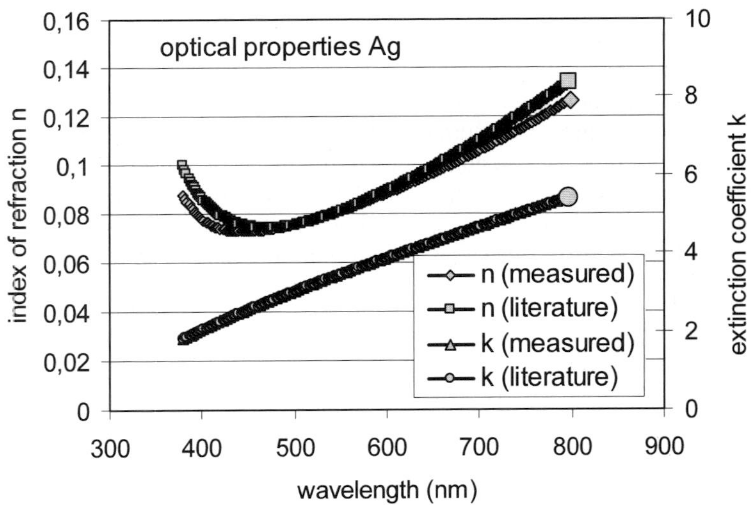

Figure 35. Index of refraction and extinction coefficient of silver as a function of the wavelength.

Finally, some measurements are given for growth-dominant (fast-growth) phase-change materials. The index of refraction and the extinction coefficient are given in Figure 36.

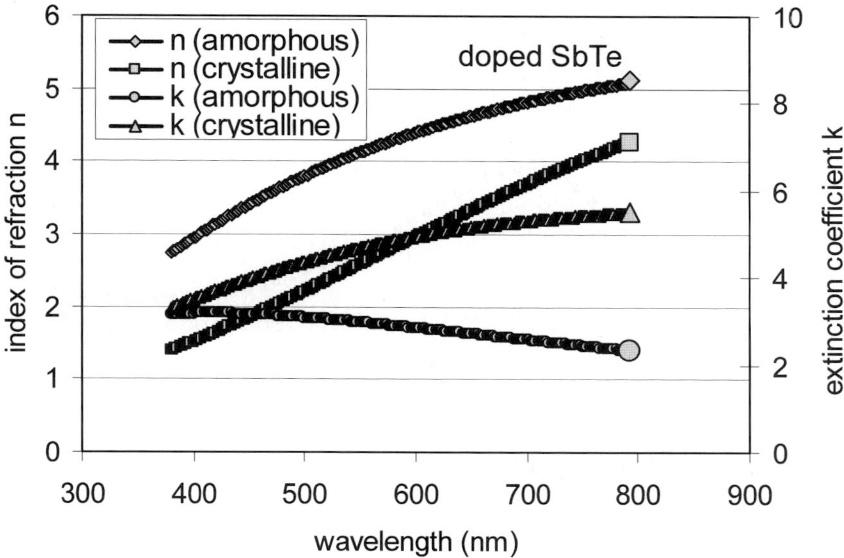

Figure 36. Index of refraction and extinction coefficient of doped SbTe phase-change material as a function of the wavelength.

3.2.2.   Stack design

Optically, a complete recording stack acts as an interference cavity, which can be fine-tuned in order to achieve best performance. Fine-tuning is done by adjusting the thickness of the layers in the stack and by a proper selection of the layer materials. To give a simple example two MI_2PI_1 stacks are considered below. The difference between the stacks is in the material of the I_1 layer, which is ZnS-SiO$_2$ in one case and SiO$_2$ in another. The recording stacks comprises further a 120 nm thick Ag M-layer, an 11 nm thick phase-change P-layer, and an 11 nm thick ZnS-SiO$_2$ second interference I$_2$-layer. In Figure 37 the calculated reflectance of the stacks and the optical contrast between the amorphous and the crystalline state are plotted versus the I$_1$-layer thickness. The measured index of refraction and absorption coefficient were input parameters for the calculation. The optical contrast is defined in this case as $C=1-Ra/Rc$, where Rc and Ra are reflectivity levels of the stack containing phase-change material in the crystalline and the amorphous state, respectively. The optical contrast is proportional to the signal modulation and, therefore, is a good measure of the optical performance of the stack. As it can be seen from the figure, the material and thickness choice have a drastic influence on the optical characteristics of the stack. In this case, an I_2 of 30 nm ZnS-SiO2 would give the highest contrast.

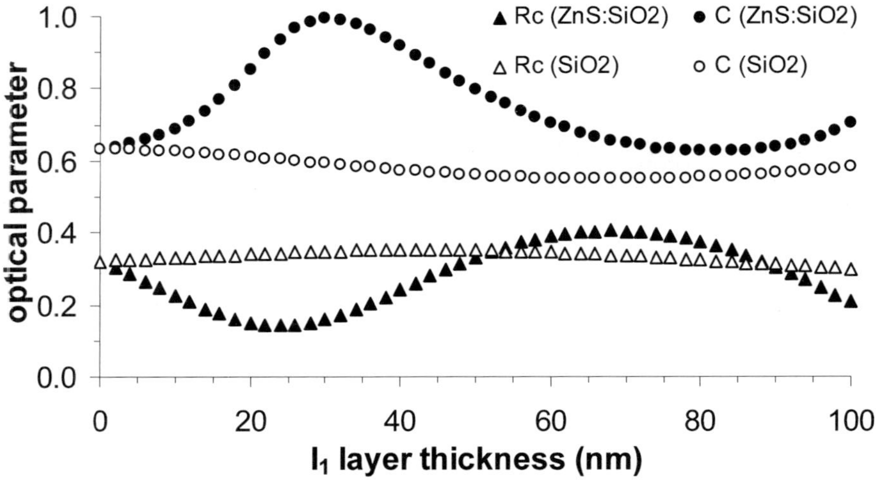

Figure 37. Optical contrast (C) and reflection of the crystalline state (Rc) as a function of the thickness of the first dielectric layer in a phase-change recording stack.

3.3.   *Thermal conductivity of thin films*

Several methods were published in the past to measure thermal properties of thin films of sub micron thickness. A good review paper was published in 2002, addressing nano-scale heat transport and listing future challenges for the measurement techniques of thermal properties of thin films. [91] In general, steady-state methods are accurate only for layers in the millimeter thickness range. The *sandwich* method, for example, is based on a sample that is sandwiched between two isothermal hotplates. The heat flux and imposed temperature gradient need to be experimentally determined to derive the thermal conductivity of the sample. [92] Transient methods, like the ac-calorimeter [93], [94], [95] and the 3ω-technique [96], [97] can be applied to accurately determine thermal properties of films of micrometer thickness. For nanometer thick films, laser-based methods provide the highest accuracy. Many laser-based methods have been published, like the thermo-reflectance method [98],

[99], the photo acoustic method [100], [101], [102] and the photo thermal wave detection method. [103], [104]

3.3.1.   *In situ laser methods*

Since phase-change recording stacks comprise several different thin films, the interfacial thermal resistances in between the films have a great influence on the effective heat transport through the stack, and need to be accounted for as well. Since it is rather hard to distinguish the interface resistance from the bulk resistance, an *in situ* approach can be followed. In that case, the measured thermal parameters allow for a good estimate of the temperature distribution during writing and erasing of data.

Mansuripur and co-workers have developed a method to determine thermal properties of amorphous phase-change recording stacks. [105], [106] The laser power required for crystallization of the amorphous phase-change layer is measured with a static tester (a laser set-up in which the sample or disc cannot spin) and is used in a numerical model. Change of the spot size results in a different thermal response and allows the determination of the thermal properties of the thin layers in combination with the measured crystallization temperature. A drawback of this method is the exact definition of the crystallization temperature since this temperature depends on the timescale of heating. Furthermore, only amorphous stacks could be tested while thermal properties of the crystalline phase are of more importance to phase-change recording because the major part of the phase-change layer is in the crystalline phase. To overcome these issues, the melt-threshold method has been proposed by Meinders *et al.* for a non-rotating disc case. [107] The method is based on the detectable reflection change upon melting of the phase-change layer. The melt temperature is not dependent on time scales of heating and properties of the crystalline state can be obtained. Another feature of this method is the use of time-resolved measurements, where the reflection of the disc is measured instantaneously rather than in a second read step after the phase change has occurred.

As an extension of the time-resolved static tester melt-threshold method, the dynamic variant has been developed. [109] A big advantage of such a method is the use of well-calibrated recorder set-ups. In addition to layer thickness and spot size variation, variation of the recording (spin) velocity of the disc can be used to change the thermal response of the recording stack. In the following section, we give a concise description of the method and some of the results achieved.

3.3.2.   *Melt-threshold method*

In this method, the reflection difference between the amorphous and crystalline states of the phase-change material is utilized to experimentally determine the laser power that causes the onset of melting of the phase-change film. In a typical melt-threshold experiment, the instantaneous disc reflection is measured as a function of the applied laser power. The sharp transition in the reflection denotes the onset of melting of the phase-change material. The maximum temperature in the recording stack is assumed to be the melting temperature. In addition, the thermal response of the phase-change stack and disc is numerically calculated with a multi-layer thermal model. All recording parameters, such as the recording velocity, the incident laser power, the actual layer thicknesses, the optical spot shape, etc. are incorporated in the thermal model to obtain an as realistic approximation of the experiment as possible. The index of refraction and absorption coefficient of all layers in the stack and substrate were measured in an independent experiment. These coefficients determine, together with the layer thickness, the total laser power absorbed in the recording stack. The absorbed laser power is the heat source term in the heat diffusion equation of the multi-layer stack. The calculated temperature distribution in the recording stack is now only determined by the thermal properties of the layers in the stack and the substrate. The specific heat of the layers and the thermal properties of the substrate can be determined in a separate experiment. If a recording stack is considered that consists of only two different materials, for example a phase-change layer that is sandwiched between two ZnS-SiO_2 dielectric layers, the problem is reduced to only two unknowns requiring two independent experiments for a unique solution.

In conclusion, the melt-temperature is used to calibrate the thermal model. Since the numerical model is a good approximation of the experiment, any of the parameters that are input to the heat diffusion equation can be solved.

3.3.2.　*Thermal conductivity results*

A sketch of the recorder that was used to determine the melt-threshold power of phase-change stacks is given in Figure 38. In the continuous heating mode, two methods were used to determine the laser power that causes the onset of melting of the phase-change layer (the melt-threshold power). 1) The reflection of the molten state is lower than that of the crystalline state. The laser power that causes the onset of melting in the center of the laser spot is then determined from the *in situ* reflection of the disc, i.e. the reflection is measured during continuous laser heating at increasing laser power. 2) In addition, the fact was utilized that the reflection of a re-crystallized molten area is higher than the reflection of the initial crystalline state. This initial crystalline state is obtained by crystallization of the as-deposited amorphous material with a low laser power since crystallization occurs at temperatures below the melt temperature. In a first write cycle, the continuous write power was incrementally increased, in a second read cycle the disc reflection was measured with a low but constant read power. The power that causes the onset of melting was derived from the sharp change in reflection.

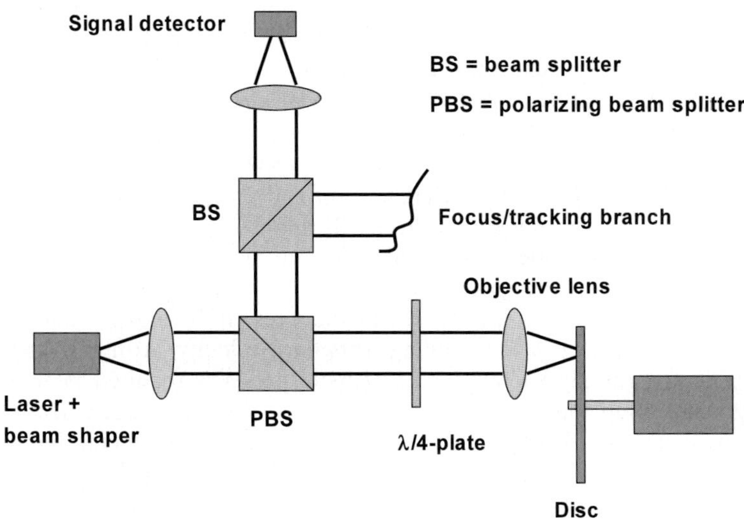

Figure 38. Sketch of the dynamic tester (recorder) used to measure the laser power that causes the onset of melting of a phase-change layer in a recording stack.

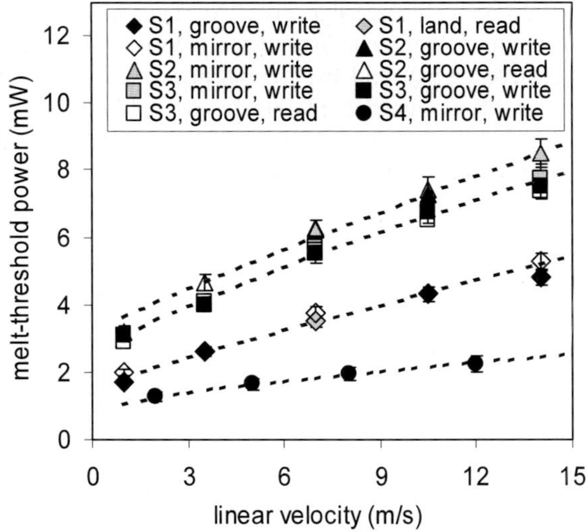

Figure 39. Measured melt threshold powers as a function of the linear velocity for four different recording stacks S_1-S_4. Stacks S_1, S_2, and S_3 were tested in a recorder with NA=0.65 and λ=658 nm, stack S_4 was tested in a recorder with NA=0.85, λ=405 nm. Shown are in-situ results obtained during continuous laser heating (indicated with 'write') and read measurements after continuous laser heating (indicated with 'read') for both the un-grooved mirror and grooved tracks (groove and land).

Measured melt-threshold powers are plotted as a function of the recording velocity in Figure 39 for 4 recording stacks, all consisting of a Sb_2Te phase-change layer that was sandwiched between dielectric layers. Stacks S_1, S_2, and S_3 were tested in a recorder with DVD optics (NA=0.65 and λ=658 nm), stack S_4 was tested in a recorder with BD optics (NA=0.85, λ=405 nm). Stack S_1 had an 8 nm Sb_2Te phase-change layer and 2 adjacent ZnS-SiO_2 dielectric layers of 90 and 140 nm thickness. Stack S_4 had a 12 nm Sb_2Te phase-change layer and 2 adjacent ZnS-SiO_2 dielectric layers of 80 and 132 nm thickness. Stacks S_2 and S_3 had additional Indium-Tin-Oxide layers. Results obtained via in-situ reflection measurements (write) and read measurements (read) for pre-grooved and ungrooved mirror parts are collected in one figure. The dotted lines are guides to the eyes. It is evident that the melt-threshold power increases with recording (spinning) velocity. At higher speeds, the dwell time is accordingly shorter such that more power is required to achieve the same maximum temperature in the recording stack. The good agreement between melt-threshold powers obtained from *in-situ* and read measurements indicate that both methods are useful and provide good reproducibility. The melt-threshold power obtained at the un-grooved (planar) regions differs slightly from that measured at the grooved regions. This difference can be understood by the difference in temperature distribution. Figure 40 gives an example of the calculated temperature distribution in the storage layer, when the laser beam is focused on a flat or grooved region of the recording stack. In the simulation, it is assumed that the grooved structure has a trapezoidal profile with track pitch 750 nm and groove depth of 40 nm. It is seen that the maximum temperature in the storage layer for land heating (the case that the laser beam is focused on a land track) is slightly lower than for groove heating (beam focused on groove track), and the temperature profile for a flat region lies somewhere in between. The presence of the groove, although only 40 nm deep, results in a somewhat deviated light absorption, and thus temperature distribution, with respect to the corresponding planar stack.

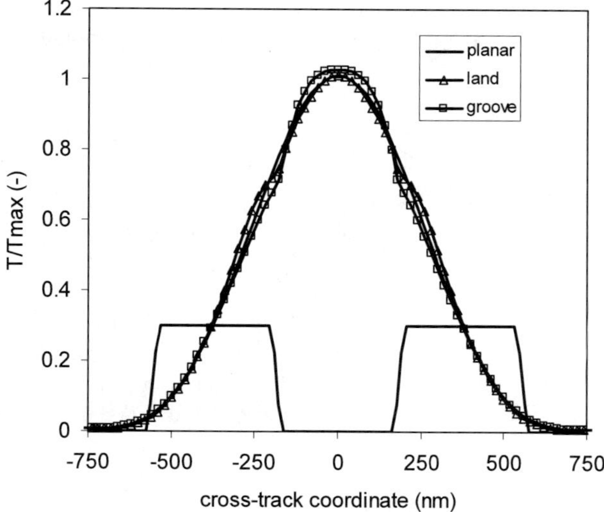

Figure 40. Calculated lateral temperature profiles for DVD conditions (NA=0.65, wavelength 658 nm, stack S_1). The light is incident from the top onto the grooved structure (solid line). Shown are profiles for a planar structure (planar) and a grooved structure (light incident on groove or land, groove depth was 40 nm, the track pitch was 750 nm). The incident laser power was the same for all three cases; the temperature is normalized with the melt temperature.

The two unknowns $\lambda_{ZnS\text{-}SiO2}$ and λ_{SbTe} in the numerical model were systematically varied to obtain curves of solutions for samples S_1 and S_4, see Figure 41. It was assumed that the thermal conductivity of the upper and the lower $ZnS\text{-}SiO_2$ layer are the same. Each set of $\lambda_{ZnS\text{-}SiO2}$ and λ_{SbTe} reproduces exactly the melting temperature when all other recording parameters are incorporated in the numerical model. The bundles of curves corresponding to stack S_1 and S_4 are clearly separated, which is caused by the different thermal response to a DVD-kind of laser spot (NA=0.65 and λ=658 nm) and a BD-kind of laser spot (NA=0.85, λ=405 nm). The sensitivity to the recording velocity is also observable but is less pronounced. The determined thermal conductivity of $ZnS\text{-}SiO_2$ and of SbTe are found at the intersection point ($\lambda_{ZnS\text{-}SiO2}$, λ_{SbTe}) = (0.92, 3.2) W/mK. The values are in agreement with the values obtained with the static variant of this method. [107] In a recent publication of Li et al, lower values are reported that have been obtained with a steady-state method. They also have measured a thickness dependence in the tested thickness range of 50-225 nm. [108]

It is also seen that λ_{SbTe} can be largely varied without involving a significant change in $\lambda_{ZnS\text{-}SiO2}$ around the cross point (note the logarithmic scale). Apparently, the steep gradient indicates that the simulated melt temperature in the phase-change layer is not very sensitive to a change in λ_{SbTe} when $\lambda_{ZnS\text{-}SiO2}$ is in the range 0.8-1.1 W/mK.

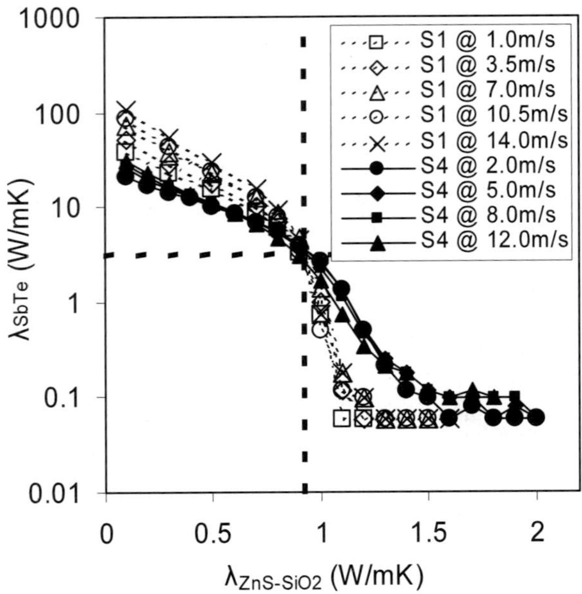

Figure 41. Simulation results for samples S_1 (NA=0.65, λ=658 nm) and S_4 (NA=0.85, λ=405 nm). The data points represent fitted combinations of λ_{SbTe} and $\lambda_{ZnS-SiO2}$ that satisfy the requirement of melting in the center of the laser spot during continuous heating for various recording velocities. The intersection point of the dotted lines represents the general solution ($\lambda_{ZnS-SiO2}$, λ_{SbTe})=(0.92, 3.2) W/mK. [109]

Table 7. Thermal conductivity (λ), heat capacity (ρc_p), index of refraction (n at 658 nm) and absorption coefficient (k at 658 nm) used in the dynamic tester simulations. [109]

	polycarbonate	SbTe (cryst.)	ZnS-SiO$_2$	Cover
λ (W/mK)	0.24	3	0.9	0.24
ρc_p (Jm^{-3}K^{-1})	1.5×10^6	1.5×10^6	2.1×10^6	1.5×10^6
(n)	1.58	3.95	2.14	1.58
(k)	0	4.25	0	0.017

3.4.   Determination of crystallization parameters

Two methods were used to determine the velocity of crystal growth. [115] Low-temperature predictions of the velocity of crystal growth were obtained from isothermal measurements performed in a temperature-controlled oven. In addition, computer simulations were fitted to time-resolved static tester measurements to obtain the temperature-dependent velocity of crystal growth $V(T)$ for the entire temperature range of interest. A static tester is basically a full-functional optical recording device with exception of the disc rotation. These simulations and experiments were performed for DVD conditions (NA=0.65, λ= 670 nm). The considered recording stack comprised four layers, the doped Sb_2Te recording layer of 10 nm thickness, sandwiched between two ZnS-SiO_2 layers of 201 and 15 nm thickness and the adjacent aluminum alloy heat sink layer of 50 nm thickness (IPIM=201-10-15-50 nm). The index of refraction and absorption coefficient of the different layers are tabulated in Table 7. The thermal properties of the layers as derived from melt-threshold experiments are also tabulated in Table 7. [107], [109]

3.4.1.   Isothermal measurements

A low-temperature prediction of the isothermal velocity of crystal growth was obtained from reflection and transmission measurements. Amorphous marks of 200 nm radius were written with a DVD recorder in a crystalline phase-change stack, which was sputter-deposited on top of a glass substrate. The samples were placed in an oven and kept at constant temperatures of 152, 157, 162 and 167°C. While kept at constant temperature, crystal growth occurred at the amorphous-crystalline interface, which finally resulted in complete erasure. The heating-up time was negligible with respect to the time for complete erasure. The degree of crystallization was derived from the time-dependent reflection measurements (see Figure 43). The different stages in the erase process can be seen in the Transmission Electron Microscopy (TEM) given in Figure 42: the initial amorphous marks (left image), the partly crystallized marks (after about 1 hour in an oven at 150°C) and the completely re-crystallized amorphous marks (right image). The images reveal that crystallization proceeds via radial crystal growth from the crystalline-amorphous interface (for this reason, the re-crystallized areas look different from the initial crystalline background). This insight was used to fit a model through the time-dependent reflection data from which the time for complete erasure was derived. The isothermal growth velocity was calculated from the time for complete erasure and initial mark radius, see Figure 44.

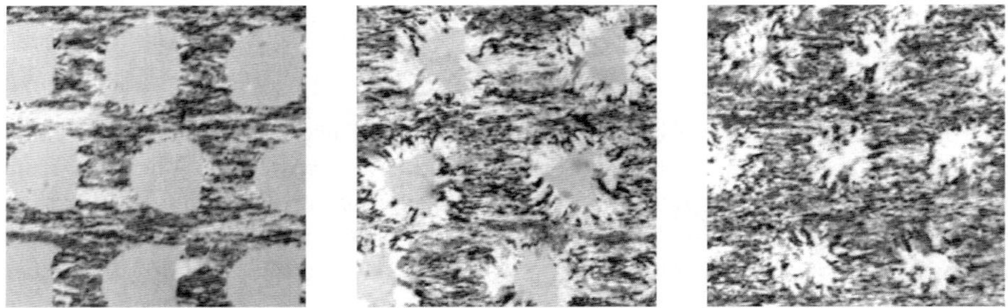

Figure 42. Transmission electron microscopy (TEM) images of written amorphous marks in a thin crystalline phase-change layer: initial state (left panel), partly erased marks after about 1 h under isothermal conditions (middle panel) and completely erased marks (right panel).

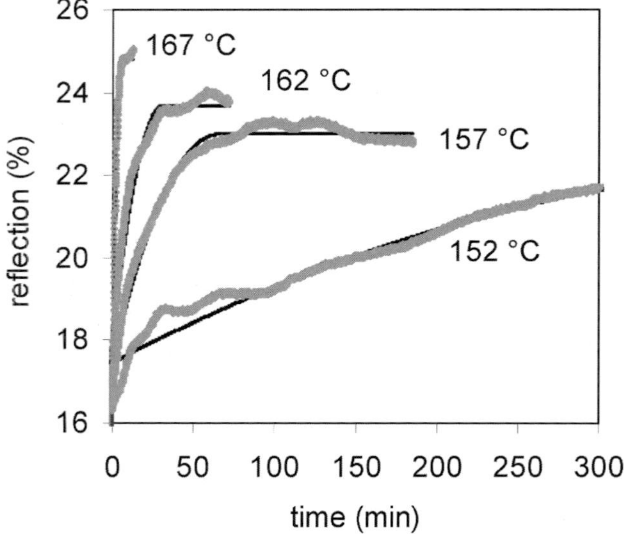

Figure 43. Time-dependent reflection of a crystalline phase-change stack with written amorphous marks for different sample temperatures. The recording stack was sputter-deposited onto a glass substrate to resist the high temperatures. The bold lines denote experiments and the solid lines denote analytical fits assuming a constant growth velocity.

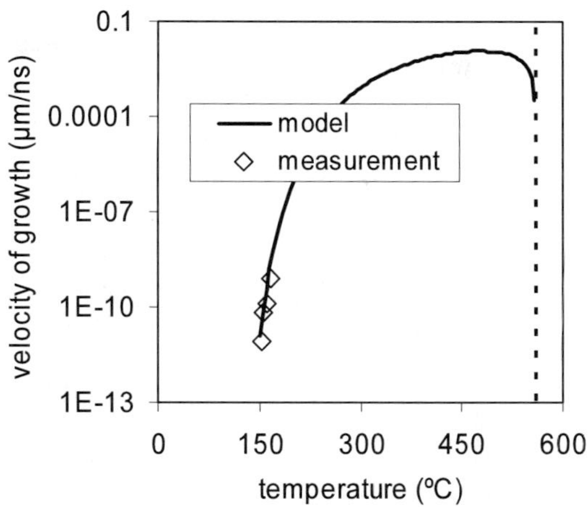

Figure 44. Measurements of the isothermal growth velocities, determined from the measured time for complete erasure and the initial mark radius. The solid line denotes the model of the corresponding temperature-dependent growth velocity (E_a=15 kJ/mol, V_0=4.5 µm/ns and T_g=358 K).

A model for the temperature-dependent velocity of crystal growth V(T) was fitted through the measured isothermal velocities, the result is given in Figure 44, with parameters T_g=358 K, V_0=4.5 µm/ns and E_a=15 kJ/mol. Although the measured velocities are far away from the temperature range of relevance for phase-change recording, namely the range in the vicinity of the melting temperature, we see a good fit. The low temperature range, in which the isothermal growth velocities were determined, is important for long-term testing, for example, for the archival life stability.

3.4.2. Time-resolved static tester measurements

The time-resolved static tester was described elsewhere. [107] In brief, a pulse generator in combination with a laser driver were used to generate laser pulses of nanosecond duration. The shortest possible pulse length was 3 ns, but typical pulse lengths of 20-100 ns were used to write amorphous marks. The time-resolved laser output and sample reflection were measured with fast photodiodes.

Amorphous marks of predefined size were written in the phase-change stack, which were subsequently erased with a long laser pulse of variable laser power. The time-resolved reflection was measured during the erase pulse and fitted against a theoretical model that describes radial crystal growth. The time for complete erasure was defined as the time required for attaining 90% of the crystalline reflection. A typical example is given in Figure 45; the derived time for complete erasure was in this case 31 ns.

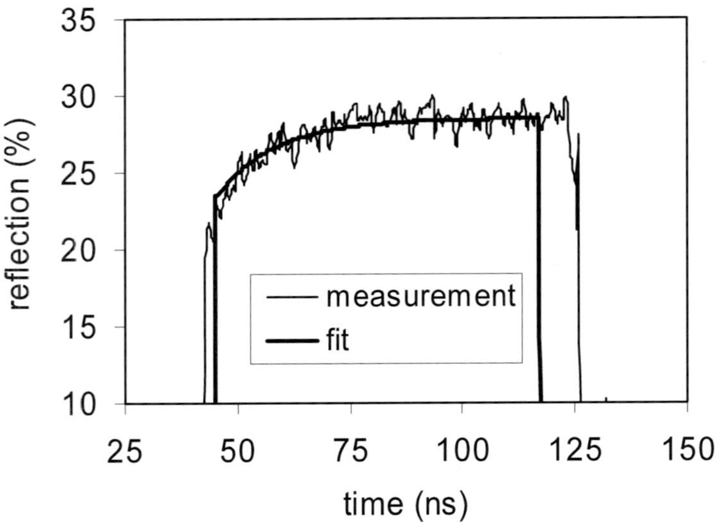

Figure 45. Measured time-resolved reflection during erasure of an amorphous mark. The time for complete erasure was defined as the time needed to attain 90% of the crystalline reflection and was derived from the fit.

The complete erasure time was determined for different erase powers, see Figure 46. Obviously, a higher erase power involves higher temperatures and thus requires less time for complete erasure. The time for complete erasure was also calculated with the numerical mark formation and erasure model. First, the response to the write pulse was calculated. The size of the written amorphous mark was determined by the initial melting of the phase-change layer and the subsequent re-crystallization during the cooling down of the mark. This amorphous mark size was input in the erasure simulation. The time needed for complete erasure was calculated as a function of the erase power for a broad range of parameters V_0 and E_a. Also, the influence of the assumed glass transition temperature T_g on the erasure behavior was investigated. From these simulations, three sets of V_0, E_a and T_g appear to provide good agreement between the measured and the calculated times for complete erasure, see Figure 46. The combinations are (E_a=7.5 kJ/mol, V_0=0.5 μm/ns and T_g=425 K), (E_a=15 kJ/mol, V_0=19.0 μm/ns and T_g=425 K) and (E_a=15 kJ/mol, V_0=4.5 μm/ns and T_g=355 K). In particular, the last set corresponds well to the isothermal growth velocity measurements as discussed before (and shown in Figure 44). The three sets of parameters result in growth curves with more or less similar growth velocities, in particular, at the lower temperature bound.

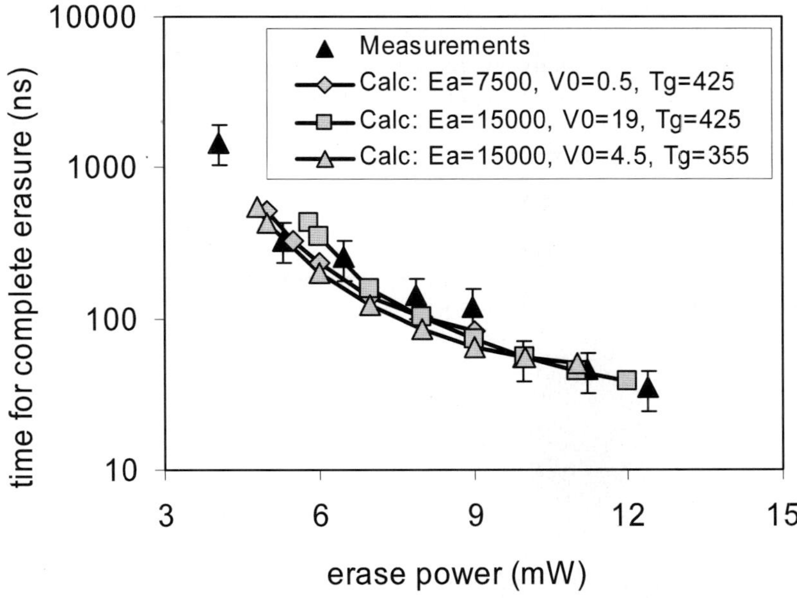

Figure 46. Comparison between measured and calculated times for complete erasure of an amorphous mark in a crystalline layer. The measurements are obtained from time-resolved static tester experiments. The calculations correspond to three sets of V_0 (µm/ns), E_a (J/mol) and T_g (K) for which the simulations were in good agreement with the experimental data.

3.5. Mark formation and erasure results

A detailed comparison between predictions and experiments was performed to validate the mark formation model.

The first example discussed is the writing of a short 2T mark in a conventional 4-layer IPIM stack under BD recording conditions (NA=0,85, 405 nm wavelength, LV=8.1 m/s). The applied laser power strategy is given in the first panel of Figure 47. An erase pulse was applied prior to and after the write pulse to simulate erasure of old data. A cooling gap in between the write and erase powers enables the control of re-crystallization. The predicted mark shape is given in Figure 47b. The solid line denotes the melt-edge; the filled gray area is the mark after re-crystallization. Since the laser spot moves from left to right, the focused spot partly intersects with the amorphous mark at the moment the erase power is activated. It is seen that post heat from this erase power causes crystal growth at the trailing edge of the amorphous mark. The temperature-dependent velocity of crystal growth explains the typical crescent shape. We recall that the temperature distribution in the direction perpendicular to the direction of the moving laser spot has a Gaussian shape. If the temperature at the central axis is around the melting temperature, the velocity of crystal growth is accordingly higher somewhat off-axis, eventually leading to a larger re-crystallized distance. A TEM picture of such a short mark is given in Figure 47c. The TEM picture reveals a clear crescent shape, illustrating the predictive power of the mark-formation model.

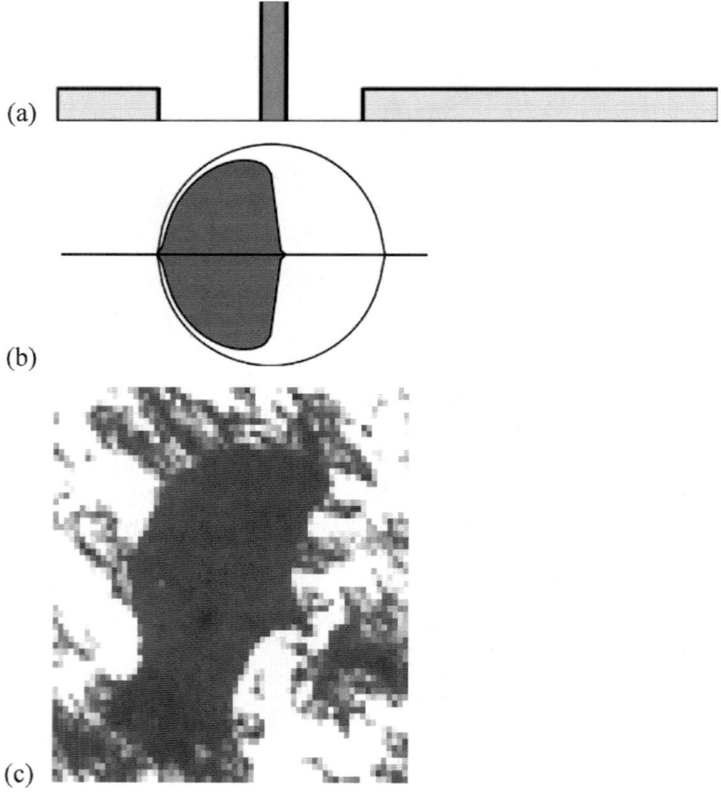

Figure 47. Simulation and recorder result of a short amorphous mark (2T) written in a DVR IPIM stack at LV=8.1 m/s. (a) Write strategy (=laser power as a function of time). (b) Predicted mark shape. The solid line indicates the molten area; the filled contour denotes the resulting mark shape after re-crystallization. (c) TEM picture of recorded mark.

A computer simulation result and a TEM image of a 7T mark are given in Figure 48. The mark was written with a so-called N-1 strategy, forming an N-channel-bit long amorphous mark. In this example, a sequence of six write pulses was applied to obtain a mark of 7 channel bits long, see Figure 48a. The computer simulation result is given in Figure 48b. The temperature rise induced by a next write pulse causes partial crystal growth of the amorphous mark written by the previous laser pulse. This partial re-crystallization leads to the serrated side edge of the mark. The temperature rise induced by the erase power at the end of the pulse train leads again to the crescent shape in the tail of the mark. The TEM picture of a 7T long mark, given in Figure 48c, reveals similar phenomena. Both the serrated edge and the re-crystallization at the trailing edge are clearly visible in the TEM picture. Again, the good agreement between the experimental and computer simulation results illustrates the predictive power of the mark formation model.

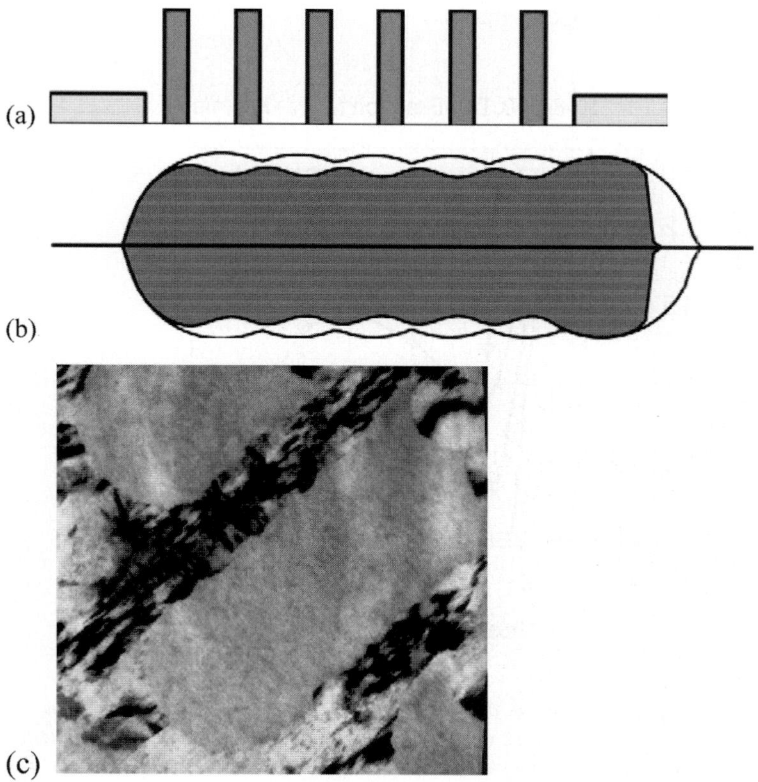

Figure 48. Simulation and recorder result of a long amorphous mark (I7) recorded in a DVR IPIM stack at LV=8.1 m/s: (a) Write strategy. (b) Predicted mark shape. The solid line indicates the molten area; the filled contour denotes the resulting mark shape after re-crystallization. (c) TEM picture of recorded mark.

Erasure of data can also be predicted with the numerical model. Long amorphous carriers were assumed to be present in an IPI recording stack. During a single passage of the laser spot, crystal growth from the amorphous mark edge was calculated for DVD conditions. The final mark width is plotted in Figure 49 for an erase velocity of 7.3 m/s. Results for three different material compositions are compared, namely a fast, a medium and a slow phase-change composition. Re-crystallization from the mark edge starts at an erase power of about 3 mW. The melting power at which the phase-change film starts to melt is indicated in Figure 49 with an arrow. For the fast material, laser powers around and above the melt-threshold result in complete re-crystallization (erasure) of the mark. For a slow composition, incomplete re-crystallization occurs whenever laser powers around the melt-threshold are applied. Application of higher erase power will not really improve the erasability but only leads to melting, and thus amorphisation, of the film. This is seen as widening of the mark. Only when an excess of heat is pumped into the stack, complete erasure will occur but the involved severe heat accumulation will deteriorate marks in the adjacent tracks.

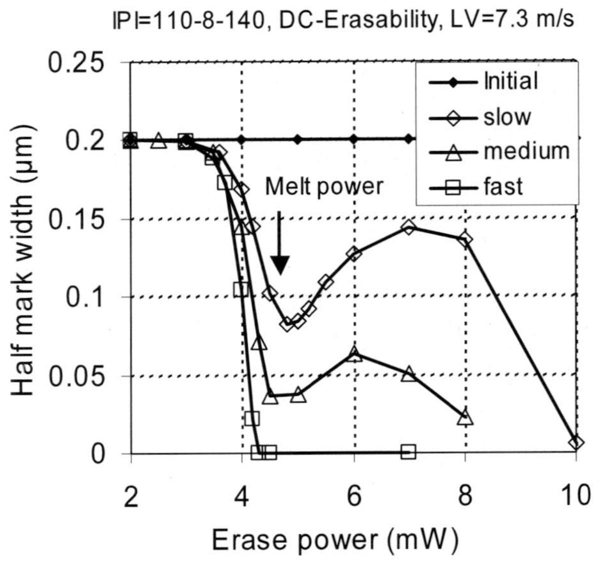

Figure 49. Erasability simulations of long marks for an IPI stack (IPI=140-8-110 nm, NA=0.65, wavelength 658 nm): half mark width after (partial) re-crystallization as a function of the erase power for three different temperature-dependent velocities of crystal growth. The linear velocity was LV=7.3 m/s. The solid diamonds indicate the initial half mark width.

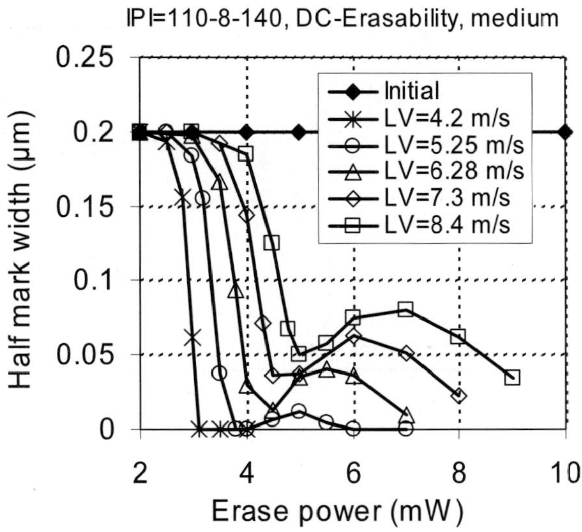

Figure 50. Erasability simulations of long marks for an IPI stack (IPI=140-8-110 nm, NA=0.65, wavelength 658 nm): half mark width after (partial) re-crystallization as a function of erase power for different linear velocities and a medium growth velocity. The solid diamonds indicate the initial half mark width.

The effect of the erase velocity on the erasability becomes clear from the simulation results given in Figure 50. In the case that the erase velocity of the laser spot is lower than the characteristic speed of crystal growth of the phase-change material, complete erasure can be obtained. This is the case for the two lowest linear speeds, namely LV=4.2 and 5.25 m/s. For the higher recording speeds, complete erasure cannot be achieved with moderate erase powers and an amorphous mark remains in the track. The cause is found in the narrower and lower temperature profile during passage of the erase spot. Further amorphisation of the track occurs if the erase power is increased to above the melting power. In that case, the crystalline spaces in between the amorphous marks become amorphous as well.

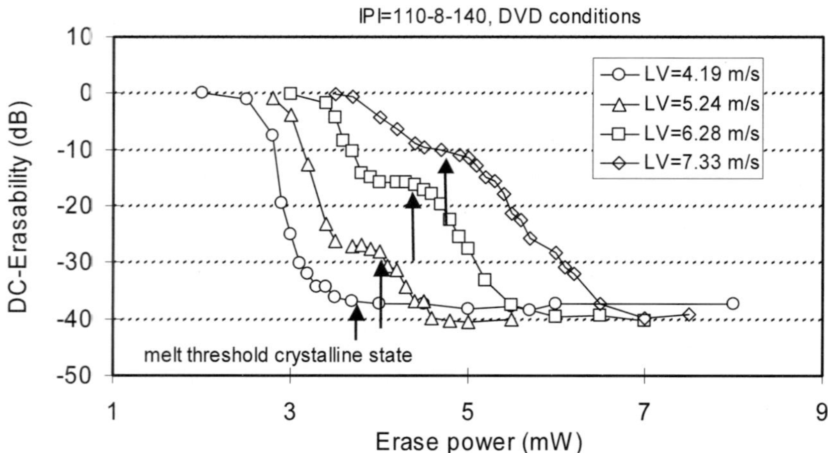

Figure 51. Measured DC-erasability of 11T carriers for the IPI stack (IPI=140-8-110nm) at DVD conditions (NA=0.65, wavelength 658 nm).

A similar erasability experiment was carried out with an IPI stack with a slow phase-change composition for different erase velocities, see Figure 51. The DC-erasability is plotted as the ^{10}log of the modulation ratio before and after erasure. The melt-threshold powers are indicated with arrows in the figure. For LV=4.2 m/s complete erasure is obtained for an erase power just below the melt-threshold power. For the higher erase velocities, a stagnation in erasability is observed just below the melt-threshold power. If the erase power is increased to above the melt threshold, both the crystalline spaces and amorphous marks melt. The combination of high quench rates, which is the case at high linear velocities, and a slow-growth phase-change composition leads to amorphisation along the entire track. This amorphisation, confirmed from a reduction in the measured reflection, leads to a reduced carrier frequency. Therefore, the measured erasability level drops further for powers above the melt-threshold power. These trends and the corresponding erase powers are in agreement with the numerical modeling results.

3.6.   *Effect of groove shape on direct heating*

3.6.1.   *Groove structure of phase-change discs*

The mechanical stability of an optical drive is insufficient to allow for an accurate positioning of the laser spot on the optical disc during recording of data. Therefore, a passive tracking system is provided, which is based on a spirally-winded pre-groove present in the optical disc. This so-called tracking is possible by light diffraction on the pre-groove structure. In brief, the −1st and 1st diffraction orders of the focused laser beam interfere with the zeroth diffraction order, the resulting pattern is received by a two-quadrant detector. In case the beam is perfectly aligned with the pre-groove, the net signal will be zero. Any shift of the laser spot with respect to the groove structure will cause a change in the diffracted pattern and will give a net signal departing from zero. This change is used in a feedback loop to re-position the spot with respect to the groove. The grooves contain a sinusoidal deflection in the radial direction. This deflection is detectable in the radial error signal (push-pull signal) and is used to synchronize the high-frequency data. Address information is encoded in discontinuities in the polarity of the wobble deflection (see also chapter 1 for a more detailed discussion on tracking mechanisms).

The groove structure has an impact on the absorbed light profile due to light diffraction at the groove walls. In particular the groove shape, like the depth, the duty cycle, the track pitch and the flank angle, affects the write performance. The groove shape has been optimized with respect to the performance of the disc in terms of push-pull signals and absorption profiles.

In Table 8 an overview of the track pitch and corresponding groove depths of three generations of optical discs is given: CD, DVD and BD.

Table 8. Three generations optical system with groove depth and data track pitch.

System	groove depth (nm)	track pitch (nm)
CD	80	1600
DVD	40	740
BD	20	320

Examples of pre-grooved BD discs are given in Figure 52. The upper picture is a cross-sectional Transmission Electron Microscopy (TEM) picture of a MIPI stack, the lower image is a cross-section TEM picture of a so-called IPI stack. The thin dark-gray layer is the phase-change layer. This layer is sandwiched between two ZnS-SiO$_2$ layers, seen as the two light-gray layers. The MIPI stack (upper image) also has an aluminum heat sink layer, the layer with the typical polycrystalline structure.

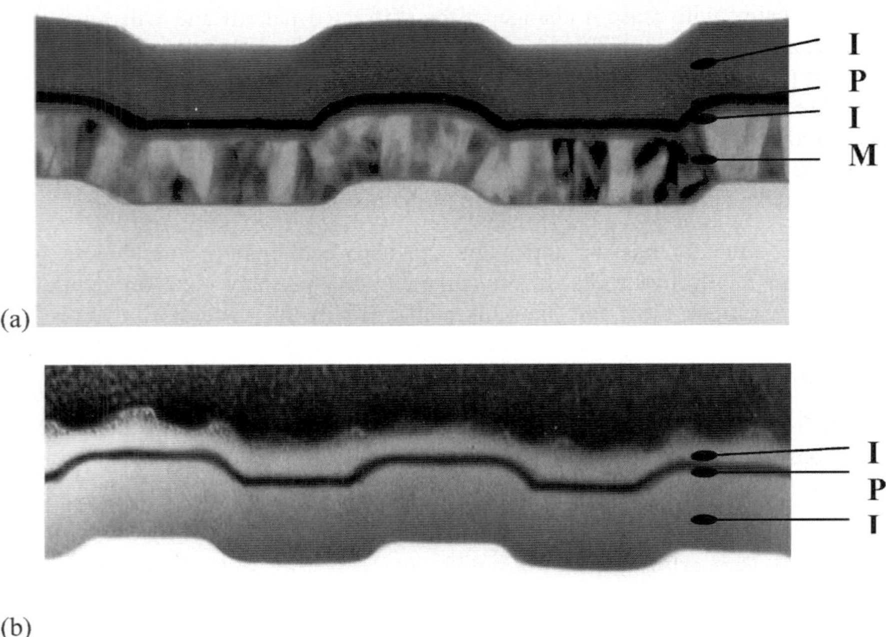

(a)

(b)

Figure 52. Cross-sectional TEM pictures of BD phase-change recording stacks. a) MIPI stack, ordered from the bottom b) IPI stack without metal heat sink layer.

To enable calculations of the light absorption in a grooved phase-change stack, the groove structure was modeled as a conformal trapezoidal structure in the XY-plane. An example is given in Figure 53. The groove structure was invariant in the Z-direction. The groove depth is indicated with d, the flank width is FW, the flank angle θ, L_1 and L_2 indicate the plateau widths and TP indicates the data track pitch. The duty cycle DC is defined as L_1/L_2. Two types of recording schemes can be considered, namely land/groove and groove-only recording. In the latter scheme, either the on-groove or the in-groove plateaus are used for data recording. This is in contrast to the formerly used land/groove recording scheme (land/groove recording was considered in the initial phase of the development of the Digital Video Recording system DVR, the predecessor of the Blu-ray Disc (BD) format, NA=0.85, λ_{laser}=405 nm) in which data were recorded on both the land and the groove plateaus. Both recording schemes are schematically indicated in Figure 54.

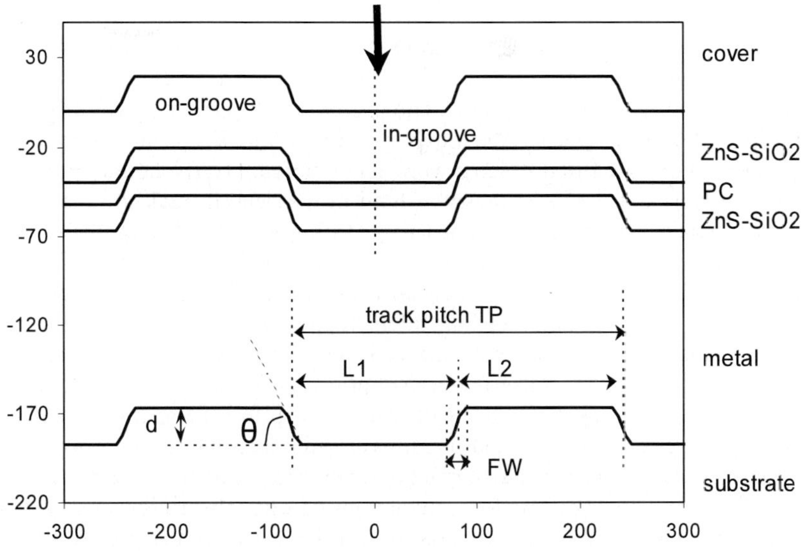

Figure 53. Schematic of the trapezoidal shaped grooved phase-change stack. The arrow indicates the direction of the incident laser beam. The different layers are indicated on the right axis.

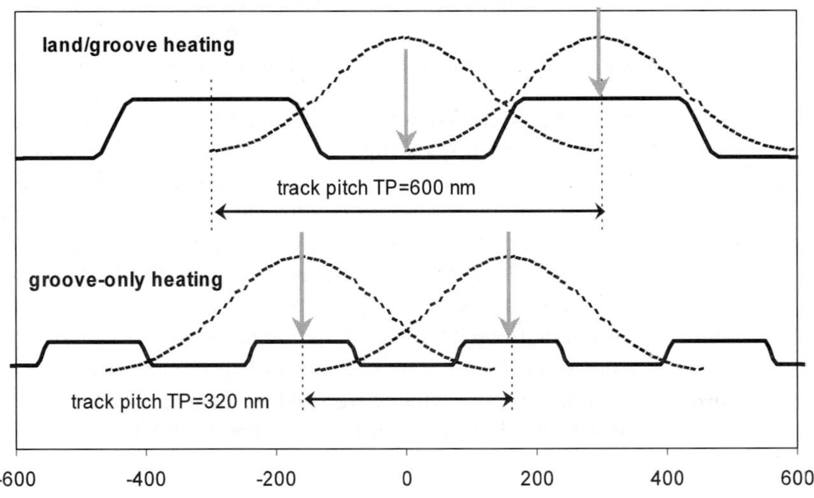

Figure 54. Schematic illustration of land/groove (TP=600nm) and groove-only heating (TP=320 nm). The solid lines indicate the groove structure; the dotted lines denote the temperature profiles.

3.6.2.   *Land versus groove recording*

Land/groove recording is characterized in that the data are written on both the land and groove plateaus of the disc (see Figure 54). In our definition, groove refers to the mastered groove in the substrate. The groove track pitch was TP=600 nm (data track pitch 300 nm). During the development of the DVR land/groove system, severe thermal cross-write problems were encountered. [116] Thermal cross-write is the phenomenon that data previously written in the central track are (partly) erased when data are written in the adjacent track. During writing of data, the sides of the thermal (and optical) spot heat the neighboring tracks such that partial erasure of marks may occur. Differences in thermal cross-write were observed between land and groove recording (land recording refers to the situation that data are written in the central land, the adjacent tracks are grooves). These differences were linked to differences between land and groove absorption. Differences in heat diffusion for land and groove heating have been discussed in. [107] Temperature calculations presented in that paper were based on an assumed Gaussian distribution of the light absorption and no rigorous vector diffraction algorithm was used. Here, temperature profiles based on vector diffraction calculations of light absorption in grooved discs are discussed. The cross-track temperature profiles for land and groove heating are given in Figure 55. Land (groove) heating denotes the situation that the spot is focused on a land (groove) plateau, the spot is then somewhat extended to the adjacent grooves (lands). The profile for land heating is 300 nm ($TP/2$) shifted to enable a detailed comparison between land and groove heating. It is seen that groove heating results in a narrower temperature distribution at the central plateau. Furthermore, groove heating is characterized by a higher temperature at the center and at the adjacent lands (side lobes). The profile for land heating is broader, has a lower maximum and shows hardly a kink at the location of the groove flank.

If the melt temperature (560°) is considered as criterion for mark formation, we clearly see that groove recording requires a higher write power in order to melt an area of the same width (to obtain a mark with the same modulation). This obviously leads to a further increase of the temperature at the adjacent lands. This temperature rise is added to the already higher temperature present at the adjacent lands (the distinct side lobes). It is clear from these temperature distributions that groove heating is more sensitive to thermal cross-erase of data present in the adjacent land. These observations are in line with recording measurements on the 'old' land/groove discs. [116] Indeed, groove recording required somewhat higher write powers. Furthermore, it was experimentally determined that groove heating caused more re-crystallization (thermal cross-write) of marks present in the adjacent land tracks.

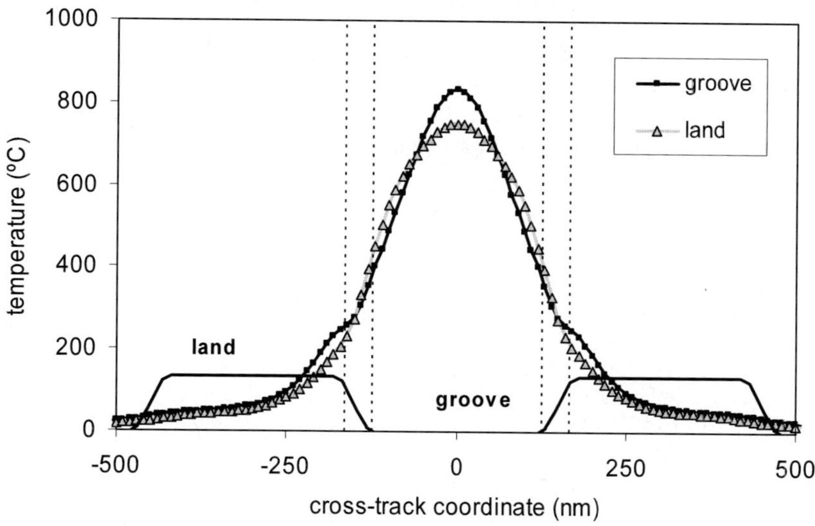

Figure 55. Cross-track temperature profiles for land and groove heating for the DVR land/groove system (NA=0.85, λ_{laser}=405 nm, TP=600 nm, flank angle 45°, groove depth d=40 nm). The land profile is shifted to facilitate the comparison between land and groove heating. The depth of the plotted groove structure (solid line) is not to scale.

3.6.3.   Groove-only recording

The groove-only recording scheme is characterized by data that are written either on the lands (in-groove) or in the grooves (on-groove) (see Figure 54). Temperature calculations were performed to illustrate the difference between in-groove and on-groove heating for a track pitch of TP=320 nm (other stack parameters are: duty cycle DC=50%, groove depth d=20 nm, flank angle 60°). Isotherm plots of the temperature distribution in the XZ-plane are shown in Figure 56 (the XZ-plane is perpendicular to the optical axis of incidence). The grooves are oriented along the Z-axis, as can be deduced from the in-plane temperature distribution. Corresponding cross-track and along-track temperature profiles are shown in Figure 57 (the on-groove profile is $TP/2$ shifted to enable a detailed comparison between in-groove and on-groove heating). Groove wall diffraction causes the clear kink in the temperature distribution for the cross-track profiles. The along-track temperature distributions at the center of the track are not visually affected by the grooved structure (the groove structure was invariant in this direction). Differences between on-groove and in-groove heating appear mainly near the groove flanks. Again, in-groove absorption (comparable to the groove case) leads to broader side lobes at the flanks walls. A major difference with land/groove recording is the absence of a significant temperature difference at the center of the track for in-groove and on-groove heating (such as seen in Figure 55).

The differences between in-groove and on-groove heating diminish far away from the center (see Figure 57). The next data track is at the location where the temperature has significantly dropped below a critical crystallization temperature. Based on the calculations, thermal cross-write seems to be irrelevant for groove-only recording.

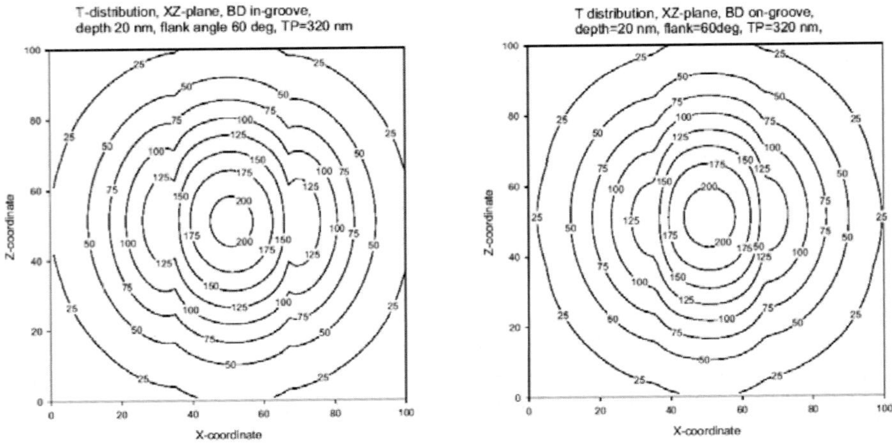

Figure 56. Isotherm representation of the temperature distribution in the plane perpendicular to the optical axis of incidence (XZ-plane) for in-groove (left plot) and on-groove (right plot) heating (TP=320 nm, d=20 nm, DC=50%, θ=60°).

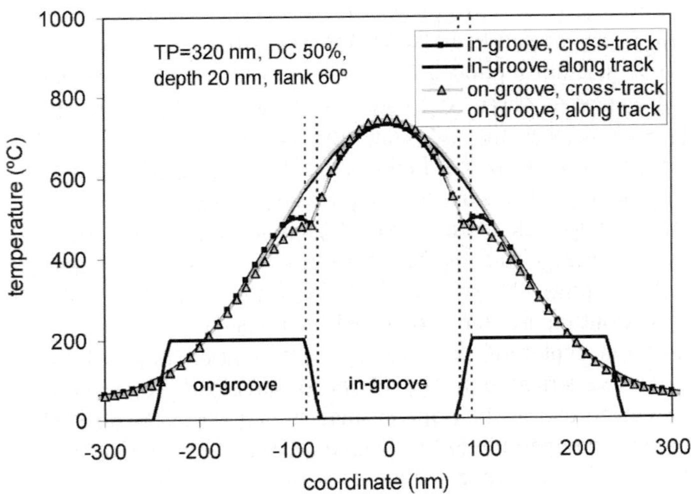

Figure 57. Cross-track temperature profiles for in-groove and on-groove heating for the Blu-ray disc groove-only system (TP=320 nm, flank angle 60°, DC=50%, groove depth d=20 nm). The depth of the plotted groove structure is not to scale.

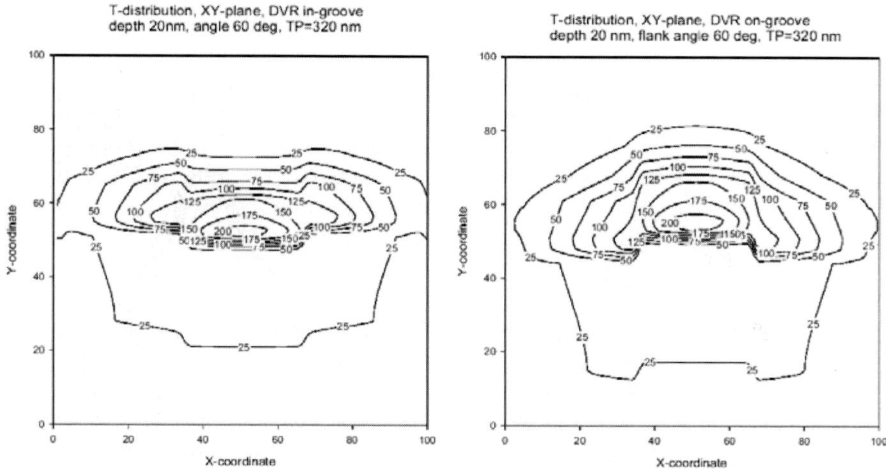

Figure 58. Isotherm representation of the temperature distribution in the cross-section plane (XY-plane) for in-groove (left plot) and on-groove (right plot) heating (TP=320 nm, d=20 nm, DC=50%, θ=60°).

Differences between in-groove and on-groove heating become also prevalent in the isotherm plots of the temperature distribution in the XY-plane, see Figure 58 (the temperatures are not scaled). It seems that the on-groove heating leads to more heat spreading in the depth (propagation) direction of the recording stack.

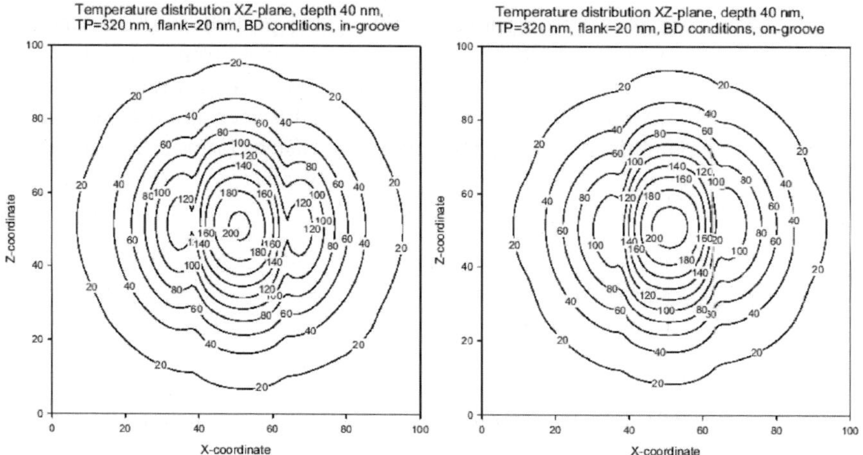

Figure 59. Isotherm representation of the temperature distribution in the plane perpendicular to the optical axis of incidence (XZ-plane) for in-groove (left plot) and on-groove (right plot) heating (TP=320 nm, d=40 nm, DC=50%, θ=60°).

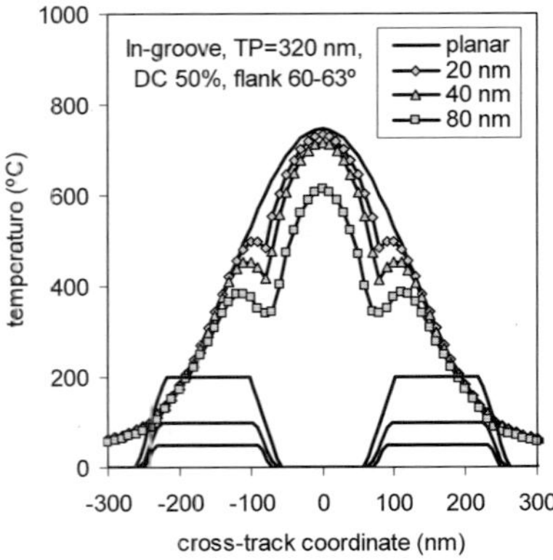
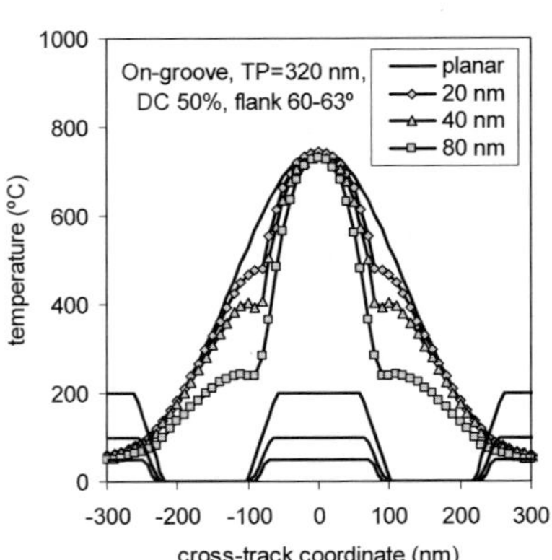

Figure 60. Cross-track temperature profiles for in-groove (left panel) and on-groove (right panel) heating as a function of the groove depth d, d=0 nm (planar), 20 nm, 40 nm and 80 nm (TP=320 nm, DC=50%, θ=60-63°). The depth of the plotted groove structure is not to scale.

Simulation results for various groove depths were performed for Blu-ray Disc conditions (DC=50%, TP=320 nm, angle θ=60-63°). Isotherm plots of the temperature distribution in the XZ-plane are given in Figure 59 for a groove depth of d=40 nm. Again, the clear side lobes are visible, being a bit sharper for the in-groove heating case. Cross-track profiles through the center of the spot are plotted in Figure 60. The planar case with zero groove depth is also plotted as a reference. As expected, the temperature profiles for shallower grooves converge towards the planar case. Obviously, the difference between in-groove and on-groove heating is most distinct for the deepest, i.e. 80 nm, groove. Such a deep groove also causes a significant reduction in the maximum temperature in case of in-groove heating.

3.7.   *References of chapter 3*

[71]   Palik, E.D., 1985, Handbook of optical constants of solids, Academic Press, San Diego.

[72]   Schijndel, M. van, Private communication.

[73]   Kittel, C., 1996, Introduction to solid state physics, John Wiley, New York.

[74]   Klitsner, T., and Pohl, R.O., 1986, Phonon scattering at a crystal surface from *in situ*-deposited thin films, Phys. Rev. B, Vol 34, pp. 6045-6048.

[75]   Klitsner, T., and Pohl, R.O., 1987, Phonon scattering at silicon crystal surfaces, Phys. Rev. B, Vol 36, pp. 6551-6565.

[76]   Guenther, K.H., 1984, Microstructure of vapour-deposited optical coatings, Appl. Opt., Vol 23, pp. 3806-3816.

[77]   Knoll, R.W., and Henager, C.H. Jr., 1992, Optical and physical properties of sputtered Si:Al:O:N films, J. Mater. Res., Vol 7, pp. 1247-1252.

[78]   C.H. Henager Jr. and W.T. Pawlewicz, 1993, Thermal conductivities of thin sputtered optical films, Appl. Opt., Vol 32, No 1, pp. 91-101.

[79]   Cattaneo, C., 1958, Sur une forme de l'equation de la chaleur eliminant le paradoxe d'une propagation instantanee, C.R. Acad. Sci., Vol 247, pp. 431-.

[80]   Vernotte, P., 1958, Les paradoxes de la theorie continue de l'equation de la chaleur, C.R. Acad. Sci., Vol 246, pp. 3154-.

[81]   Guillemet, P., Bardon, J.P., and Rauch, C., 1997, Experimental route to heat conduction beyond the Fourier equation, Int. J. Heat Mass Transfer, Vol 40, No 17, pp. 4043-4053.

[82]   Guo, Z.-Y., and Xu, Y.-S., 1995, Non-Fourier heat conduction in IC chip, J. Electronic Packaging, Vol 117, pp. 174-177.

[83]   Bai, C., and Lavine, A.S., 1995, On hyperbolic heat conduction and the second law of thermodynamics, J. Heat Transfer, Vol 117, pp. 256-263.

[84]   Ozisik, M.N., and Tzou, D.Y., 1994, On the wave theory in heat conduction, J. Heat Transfer, Vol 116, pp. 526-535.

[85]   Carslaw, H.S., and Jaeger, J.C., 1959, Conduction of heat in solids, Oxford University press, Oxford.

[86]   E.R. Meinders, H.J. Borg, M.H.R. Lankhorst, J. Hellmig, and A.V. Mijiritskii, 2002, Numerical simulation of mark formation in dual-stack phase-change recording, J. Appl. Phys., 92, No 12 (2002) pp. 9794-9802.

[87]   M. Mansuripur, G.A.N. Connell and J.W. Goodman, 1982, 'Laser-induced local heating of multilayers', Appl. Opt., Vol. 21 No 6, pp. 1106-1114.

[88]   M. Mansuripur and G.A.N. Connell, 1983, 'Laser-induced local heating of moving multilayer media', Appl. Opt., Vol. 22 No 5, pp. 666-670.

[89]   C. Peng and M. Mansuripur, 2000, Thermal cross-track cross talk in phase-change optical disk data storage, J. Appl. Phys., Vol 88, No 3, pp. 1214-1220.

[90]   Y. Nishi, J. Ushiyama., H. Kando, M. Terao, Phase change mark simulator for Ag-InSbTe recording media, Proceedings ISOM 2000, pp. 36-37 (2000).

[91]   D.G. Cahill, K. Goodson, and A. Majumdar, 2002, Thermometry and thermal transport in micro/nanoscale solid-state devices and structures, Trans ASME-C, J. Heat Transfer, Vol 124, pp. 223-241.

[92]   Maglic, K.D., Cezairliyan, A., Peletsky, V.E., editors, 1984, Compendium of Thermophysical property measurement methods, Volume 1, Survey of measurement techniques, Plenum Press, New York.

[93]   Hatta, I., 1985, Thermal diffusivity measurement of thin films by means of an ac calorimetric method, Rev. Sci. Instrum., Vol 56, No 8, pp. 1643-1647.

[94]   Gu, Y., and Hatta, I., 1991, Development of ac calorimetric method for thermal diffusivity measurement IV: films with low thermal diffusivity and very thin films, Jpn. J. Appl. Phys., Vol 30, No 6, pp. 1295-1298.

[95]   Tye, R.P., and Maesono, A., 1993, Characterisation and thermal properties evaluation of thin films, wafers and substrates, Thermochim. Acta, Vol 218, pp. 155-172.

[96]   Cahill, D.G., Fischer, H.E., Klitsner, T., Swartz, E.T., and Pohl, R.O., 1989, Thermal conductivity of thin films: measurement and understanding, J. Vac. Sci. Technol. A., Vol 7, No 3, pp. 1259-1266.

[97]   Cahill, D.G., Katiyar, M., and Abelson, J.R., 1994, Thermal properties of a-Si:H thin films, Phys. Rev. B, Vol 50, No 9, pp. 6077-6081.

[98]   Paddock, C.A., and Eesley, G.L., 1986, Transient thermoreflectance from thin metal films, J. Appl. Phys., Vol 60, No 1, pp. 285-290.

[99]   Claeys, W., Dilhaire, S., Schaub, E., 1997, Laser probing techniques and methods for the thermal characterisation of microelectronic components, Proceedings, EUROTHERM, seminar no 58, pp. 227-237, Nantes, France.

[100]   Rosencwaig, A., and Gersho, A., 1976, Theory of the photoacoustic effect with solids, J. Appl. Phys., Vol 47, No 1, pp. 64-69.

[101]   Mandelis, A., Schoubs, E., Peralta, S.B., and Thoen, J., 1991, Quantitative photoacoustic depth profile profilometry of magnetic field-induced thermal diffusivity inhomogeneity in the liquid crystal octylcyanobiphenyl, J. Appl. Phys., Vol 70, No 3, pp. 1771-1777.

[102]   Mandelis, A., Peralta, S.B, and Thoen, J., 1991, Photoacoustic frequency-domain depth profiling of continuously inhomogeneous condensed phases: Theory and simulations for the inverse problem, J. Appl. Phys., Vol 70, No 3, pp. 1761-1770.

[103]   Lan, T.T.N., Seidel, U., and Walther, H.G., 1995, Theory of microstructural depth profiling by photothermal measurements, J. Appl. Phys., Vol 77, No 9, pp. 4739-4745.

[104]   Lan, T.T.N., Seidel, U., Walther, H.G., Goch, G., Schmitz, B., 1995, Experimental results of photothermal microstructural depth profiling, J. Appl. Phys., Vol 78, No 6, pp. 4108-4111.

[105]   Y.-C. Hsieh, M. Mansuripur, J. Volkmer and A. Brewen, 1997, Measurement of the thermal coefficients of non-reversible phase-change optical recording films, Appl. Opt., Vol. 36 No 4, pp. 866-872.

[106]   C. Peng and M. Mansuripur, 1999, Measurement of thermal conductivity of erasable phase-change optical recording films, Appl. Opt. Vol. 39 No 14, pp. 2347-2352.

[107]   E.R. Meinders, M. Lankhorst, H.J. Borg, and M.J. Dekker, 2001, Thermal cross-erase issues in high-data-density phase-change recording, Jpn. J. Appl. Phys., Vol 40, pp. 1558-1564.

[108]   C.T. Li, Y. Yang, S. M. Sadeghipour and M Asheghi, Thermal conductivity measurements of the ZnS-SiO$_2$ dielectric films for optical data storage applications, 2004, ASME proceedings of IMECE04, IMECE2004-62150, pp. 225-229.

[109]	E.R. Meinders and C. Peng, 2003, Melt-threshold method to determine the thermal conductivity of thin films in phase-change optical recording stacks, J. Appl. Phys., 93 issue 6, pp. 3207-3213.

[110]	H. Kando, Y. Nishi, M. Terao, T. Maeda, M Miyamoto, A. Hirotsune, J. Ushiyama, Simulation on phase-change mark forming process and read out signal characteristics, Technical Digest ODS 2000, pp. 50-52 (2000).

[111]	M. Horie, T. Ohno, N, Nobukuni, K. Kiyono, T. Hashizume, and M. Mizuno, 2002, Material characterisation and application of eutectic SbTe based phase-change optical recording media, in Optical Data Storage 2001, SPIE 4342, pp. 76-87.

[112]	A.C. Sheila, T.E. Schlesinger, 2001, Simulation studies of mark formation and read signals in growth dominant phase-change optical recording media, Technical Digest, Optical Data Storage 2001, pp. 211-212.

[113]	J.W. Christian, 1965, The theory of transformations in metals and alloys, Pergamon Press, Oxford.

[114]	L. Battezatti and A.L. Greer, 1989, The viscosity of liquid metals and alloys, Acta metall. 37 No. 7, pp. 1791-1802.

[115]	E.R. Meinders and M.H.R. Lankhorst, 2003, Determination of the crystallisation kinetics of fast-growth phase-change materials for mark-formation prediction, Jpn. J. Appl. Phys., Vol 42, Part 1, No 2B, pp. 809-812.

[116]	M.J. Dekker, N. Pfeffer, M. Kuijper, I.P.D. Ubbens, W.M.J. Coene, E.R. Meinders, and H.J. Borg, 2000, Blue phase-change recording at high data-density and data rates, SPIE 4090, pp. 28-35.

4. Data recording characteristics

4.1. Data recording

4.1.1. Recording, erasing and direct-overwriting of marks

In optical data storage user information is encoded in the length of the written marks. Therefore, to store data onto a phase-change disc amorphous marks of certain lengths have to be formed in the recording layer. Different run-length modulation codes are known. In the DVD format, the EFMplus code is prescribed, containing marks and intermediate spaces of length between 3T and 11T, T being the channel bit length. For Blu-ray Disc, the 17PP code is defined, with mark lengths between 2T and 8T. In conventional optical drives, a spiral of marks is written on a rotating disc by applying laser pulses of certain power. In the simplest case, the length of the written mark is proportional to the total on-time of the laser in case the disc rotates at a constant linear velocity. Unfortunately, such a simple write strategy with a block-shape write pulse does not work well in the case of phase-change media. The longer the (high-level) write power is "ON" the more heat is accumulated in the disc and the longer it takes to cool down the stack. In fact, it is virtually impossible to write a long mark with a block-shaped laser pulse because the severe heat accumulation leads to re-crystallization of the mark during writing.

As has been discussed in chapter 3, heat accumulation can be managed by the thermal design of the recording stack. However, when the stack design accounts for all optical, thermal, chemical, and mechanical requirements, it might still not be possible to achieve sufficient cooling rate.

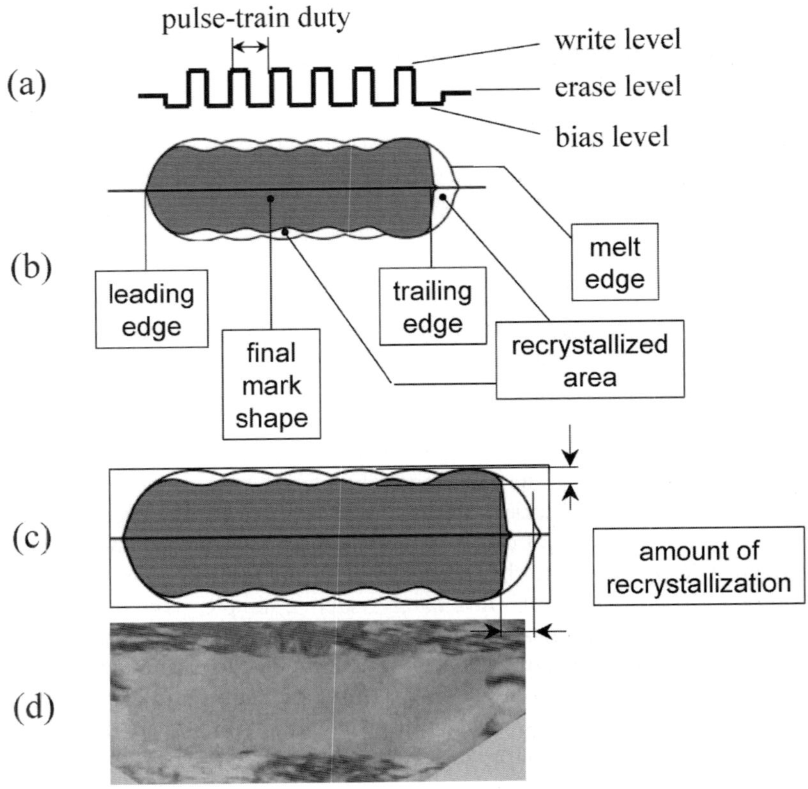

Figure 61. (a) laser power profile for writing a long amorphous mark in a phase-change layer (b) modeling result of an amorphous mark illustrating the molten area and the final amorphous area after re-crystallization c) amount of re-crystallization (d) Transmission Electron Microscopy picture of a mark.

The problem of unwanted heat accumulation during writing is usually solved by means of pulsed write strategies. In such write strategies mark writing is realized with a sequence of short write pulses separated by cooling gaps. Such a sequence is often called a pulse-train. In panel (a) of Figure 61 a laser-power profile of a typical pulse-train is shown. During each write pulse a molten dot is formed in the phase-change layer. The cooling gaps with only a very low laser power (the bias level) cause fast cooling down of the phase-change layer to enable melt quenching. To create an amorphous mark with a length longer than the spot size, a number of consecutive write pulses are closely placed such that these pulses cause overlaying molten areas. As a result, the mark formed consists of overlapping amorphous dots.

Mark formation occurs at relatively high recording temperatures of 600-800°C since melting of the phase-change layer is involved. The formation of amorphous material is simply realized by exceeding the melt-temperature of the material and can be considered as a threshold effect. Erasure of data requires a totally different temperature regime. Erasure of amorphous marks is enabled by re-crystallization, which occurs at much lower temperatures, typically around 250-400°C. The degree of crystallization is determined by the temperature-time history of the phase-change film. Erasure of marks is typically done with a continuous pulse of a moderate laser power level.

To speed up the recording process and, ultimately, to increase the user data rate, optical drives are designed to write new data over the old data within one passage of the laser spot. This method is called direct overwrite (DOW) and is realized by combining write-pulse trains and erase pulses within one write strategy. In such a case new data marks are written over the old data marks and spaces whereas the erase pulses create new data spaces by erasing the remaining pieces of the old amorphous data marks. Direct overwriting requires dedicated materials and write strategies that enable re-crystallization of old marks while maintaining sufficiently high cooling rates of the recording stack during recording of data. One of the main challenges of phase-change recording is the optimization of material properties and write strategies for specific recording speeds and applications (for example a dual-layer disc). Material optimization is typically focused on finding a phase-change alloy with a suitable crystallization speed and optical contrast, dielectric materials to enhance the contrast and crystallization, and metals to provide high quenching rates. Developing write strategies is typically done in a two-dimensional domain with laser power levels and pulse durations as variables.

The process of phase-change data recording is discussed in the next paragraph in more detail.

4.1.2.   Pulse trains

In panel (d) of Figure 61 a TEM image of an amorphous mark is given. The mark is written using the pulsed DOW write strategy shown in the upper panel of Figure 61. As can be seen from this image the shape of a real mark is rich with features. These features have a drastic influence on the modulation profile and, ultimately, are vital for robust bit detection. To understand the nature of all these features and gain more insight into the recording process, computer simulations have proven to be indispensable. In Figure 61 a result of a computer-simulated mark is given (see panels (b) and (c)). The solid line in the figure indicates the edge of the molten area upon writing. The filled area represents the final mark shape. As can be seen by comparing the real and simulated marks all of the features of the mark shape are well reproduced by the simulations.

Creating a mark in the recording layer is an inter-play between amorphisation and crystallization of the phase-change material. The write pulses determine the amount of material being molten in the recording layer. Their power level and duration are primarily responsible for the width of the amorphous mark being written. The width of the mark, in turn, determines the amount of reflection modulation and, thus, the signal quality upon readout.

The bias-power level of the cooling gaps in a pulsed write strategy is usually kept as low as possible to ensure a maximum quenching rate. The minimum bias power is limited by the stability in the laser operation and is typically chosen just above the lasing threshold of the laser. The length of the cooling gaps separating the write pulses controls the quenching rates and, therefore, the amount of material re-crystallization.

To create a continuous mark, the time between consecutive write pulses is chosen such that the molten areas partly overlap. The temperature raise induced by a next write pulse causes crystal growth at the mark edge of an earlier written amorphous dot. Thus, a sequence of alternating write and bias levels accompanied by such re-crystallization results in the serrated side edges of the mark.

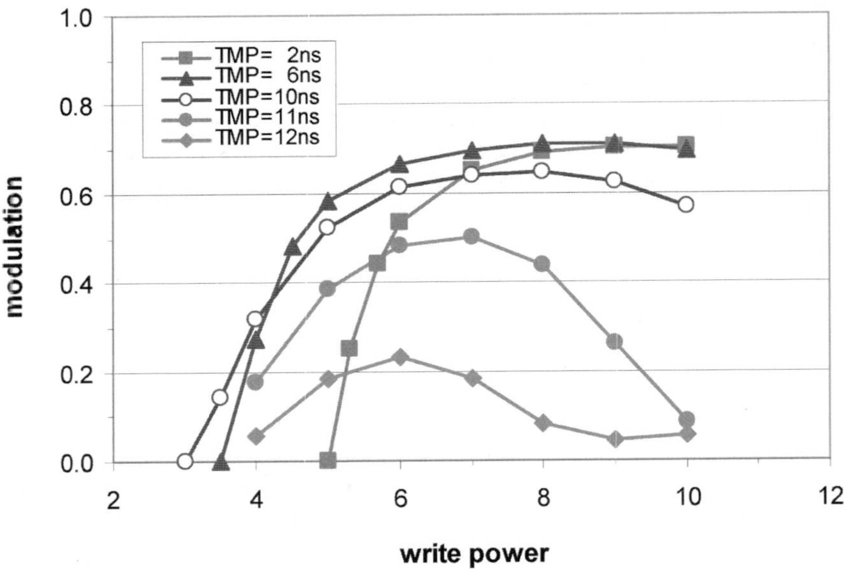

Figure 62. 14T carrier modulation as a function of laser power for various pulse duty cycles (expressed in the length of the multi pulse Tmp).

In Figure 62 a typical result of a modulation test is shown. In this example a 14T carrier was written in a BD disc with different write powers and pulse-train duty. A 14T carrier means in this context a repeating signal of alternating marks and spaces of 14T length. The length of a pulse period (write pulse plus bias) in the pulse train is 15 ns in this case. The write-pulse length of 2, 6, 10, 11, and 12 ns correspond to the pulse-train duty of 0.13, 0.40, 0.66, 0.73, and 0.79, respectively. The modulation is measured as the difference between the highest and the lowest reflectance level normalized by the highest reflectance. The general modulation profile for a given pulse-train duty is explained as follows. When the write power is too low no material is molten and no amorphous mark is formed in the recording layer. This results in a zero signal modulation since only the initial crystalline background material is present. When the write power exceeds the melt-threshold power and the quenching rate is sufficiently high, a narrow mark is formed. This is seen as an onset in the signal modulation and is a good indicator of the media sensitivity with respect to write power. A further increase in the write power results in larger molten areas and broader final marks giving rise to a higher modulation. Starting from a certain write-power level saturation in the modulation is observed. From this point on dissipation of the heat becomes insufficient and any further increase in the molten area is followed by a subsequent re-crystallization of the same area. When the write power is too high the quenching process starts to fail and a re-crystallization-caused drop in modulation is observed. In the case of low pulse-train duties (2 ns write-pulse in Figure 62) the energy dissipated in the stack is low and the cooling rate is high. At these conditions, a relatively high write power (> 5 mW) is required to create an amorphous mark and no significant re-crystallization is observed even at powers up to 10 mW. By increasing the pulse-train duty less power is required to achieve an on-set in modulation. If the duty is too high (10, 11, 12 ns write pulses in the figure) the cooling rate controlled by the cooling bias gaps becomes insufficient. The cooling failure boosts severe re-crystallization with the consequence that no broad marks and, therefore, no high modulation can be written in the recording stack. Furthermore, the power range within which the modulation stays in saturation becomes narrow. The modulation test described above is one of the basic elements of pulse-train optimization. One of the goals of such an optimization is to come up with a pulse-train that provides maximum modulation at reasonable powers and a broad enough power margin.

4.1.3.   Erase of marks

Erasability is measured as a reduction in the CNR of a data carrier. Usually, erasability is considered to be sufficient (complete erasure) if the data signal is reduced by 25-30 dB. In this case no trace of the previous data pattern is detected by the drive. In a typical erasability experiment a data carrier (preferably of the maximum bit-length e.g. 8T in the case of BD) is written and subsequently erased and the CNR difference before and after the erasure cycle is measured. To obtain a complete overview the measurements are often done at various erase powers and various disc velocities. In Figure 63 results of typical erasability tests are given. The measurements were performed on semi-transparent BD stacks (see Chapter 5) having an IMIPI-type stack design with a thin Ag mirror. The upper panel corresponds to a faster growth-dominated material, the lower panel to a slow composition. The general trend in the erasability behavior is quite obvious. If the erase power is chosen too low the temperature reached in the recording layer remains below the crystallization temperature and no erasure takes places. This leads to no measurable decrease in the signal amplitude of the present marks and therefore in a zero erasability. With increasing erase power, the partial re-crystallization of amorphous marks sets in. This is detected as a decrease in CNR. Furthermore, for a higher disc velocity the dwell time of the laser spot is shorter. As a consequence, the temperature reached in the stack is lower and more power is needed to achieve erasure. This explains the shift in erase power for a higher erase velocity. A phase-change material with a high velocity of crystal growth requires less time for complete erasure. This explains the larger CNR reduction for the faster FGM-type material (shown in upper panel in Figure 63) for similar experimental conditions, such as dwell time (erase velocity) and erase power. It is noticed that the optical characteristics, in particular the laser light absorption, and the thermal behavior of the recording stacks with fast FGM-type material (upper panel in Figure 63) and slow FGM-type material (lower panel in Figure 63) were similar to allow a fair comparison of the CNR reduction curves.

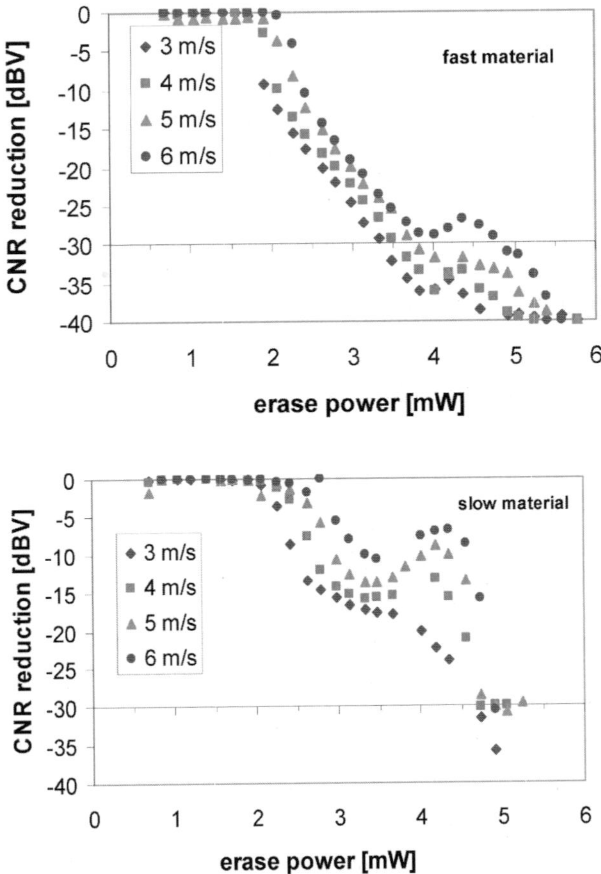

Figure 63. CNR reduction of an 8T carrier as a function of erase power for various erase velocities for a faster FGM-type material (upper panel) and a slower FGM-type material (lower panel).

Good understanding of the erasability curve profiles can be gained from computer simulation results. For this purpose the erasability of marks written in an FGM-based IPI stack was simulated with the crystallization model as described in chapter 3. Long amorphous carriers with an initial width of 200 and 400 nm were assumed to be present in the phase-change layer. The simulations with 200 nm wide marks refer to BD conditions; the 400 nm wide marks refer to DVD conditions. During a single passage of the laser spot, re-crystallization from the amorphous mark edge was calculated. Complete or incomplete erasure was obtained, depending on the linear velocity, the crystal growth velocity and erase power.

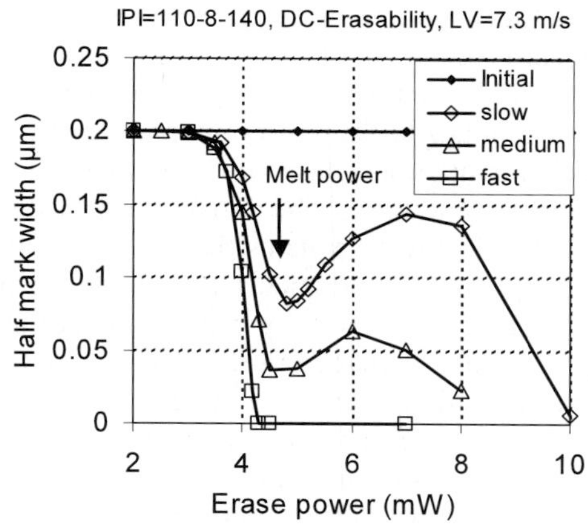

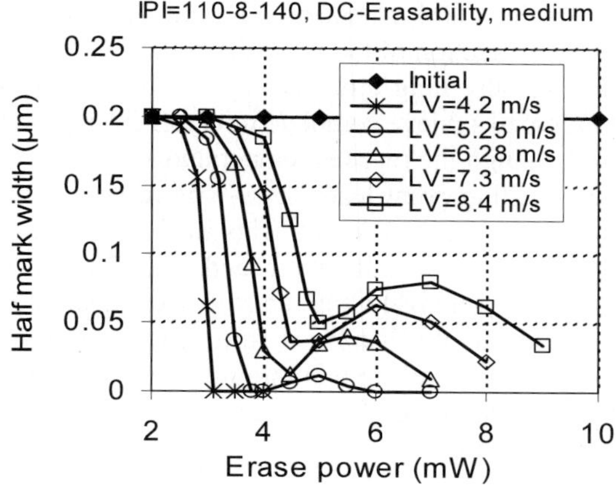

Figure 64. Simulation results of re-crystallization of amorphous marks in a slow-cooling IPI stack for DVD conditions (IPI=140-8-110 nm, NA=0.65, λ=658 nm): (a) resulting half mark width as a function of erase power for three temperature-dependent velocities of crystal growth. The erase velocity was LV=7.3 m/s. (b) resulting half mark width as a function of the erase power for different erase velocities. The material had a medium velocity of crystal growth. The solid diamonds indicate the initial half mark width.

Simulation results for DVD conditions and a slow-cooling IPI stack are given in Figure 64. The erase velocity was 7.3 m/s. The procedure of calculating the final mark width was as follows. A long mark with initial width of 400 nm was assumed to be present in the phase-change layer. The temperature response to continuous laser heating (erasure) was calculated and used as input to calculate the temporal crystal growth from the crystalline-amorphous interface during a single passage of the laser spot. The final amorphous mark shape was obtained after complete passage of the laser spot. These calculated mark widths are plotted as a function of the applied continuous erase power in the upper Figure 64 (a) for three different types of FGM materials: a material with a low, medium and high velocity of crystal growth. The temperature-dependent velocities of crystal growth were determined according to the analysis presented in chapter 3, the average velocities were 5, 8 and 10 m/s for the slow, medium and fast FGM-type phase-change composition, respectively. Re-crystallization from the initial amorphous mark edge starts at an erase power of about 3 mW, which corresponds to the onset temperature of crystal growth. For the fast FGM-type material, a continuous erase power of about 4 mW leads to complete re-crystallization (erasure) of the mark. The non-linear behavior of the temperature-dependent velocity of crystal growth explains the observed high sensitivity of mark width reduction with respect to the applied laser power. In particular in the low temperature regime, the velocity of crystal growth possesses an exponential behavior. Onset of melting of the phase-change film starts at a continuous laser power of 4.6 mW. The velocity of crystal growth of the fast FGM-type material is sufficiently high to ensure complete erasure prior to melting of the phase-change layer. Increasing the laser power to beyond the melt-threshold power leads to melting of the phase-change film, but the molten material re-crystallizes in the cooling down period after passage of the laser spot.

This is in contrast to the FGM compositions with medium and low velocity of crystal growth. Re-crystallization of the amorphous mark proceeds much slower and results in incomplete erasure long after passage of the laser spot. The velocity of crystal growth is too low to achieve complete re-crystallization in the available dwell time of the laser spot, though the reached temperatures are quite high. Increasing the power to beyond the melting point leads to re-writing of the marks by the erasing beam, and a residual amorphous mark remains in the phase-change layer. This is observed as a stagnation in the mark width and, therefore, in erasability, see the upper plot in Figure 64. At relatively high erase powers, the amount of accumulated heat in combination with the dwell time is sufficient again to induce complete re-crystallization. This behavior leads to a bump in the mark width curve. Such bumps are also observed in the measured CNR reduction curves given in Figure 65, even for the fast FGM composition.

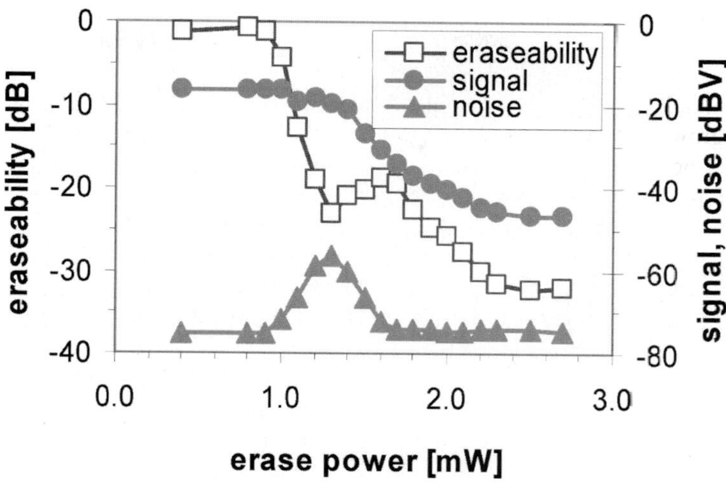

Figure 65. Erasability results of an IPIM stack.

For these experimental results, two additional phenomena are believed to contribute to the bump-like profile. One is the inhomogeneous erasure, which results in a sudden rise in the media noise level. This noise increase leads to a quicker fall of the CNR than it could be expected on the basis of modulation reduction. In Figure 65 an erasability curve is shown together with the corresponding signal and noise profiles after erasure. As can be seen from the figure the signal reduces monotonously with increasing erase power whereas a significant increase in the noise level occurs for powers close to the melt-threshold power. Such behavior causes an additional drop in CNR of the erased marks next to the decrease caused by the signal drop. Due to the presence of crystallites in the phase-change layer as well as variation in the temperature reached along the track caused by jitter in tracking, relatively large variations in crystallization occur both within a mark and from mark to mark. In Figure 66 modulation profiles of a BD 8T single-tone carrier before erasure (upper trace) and after erasure at 1.4 mW erase power (lower trace) are given. As can be seen, erasure with such conditions results in a very large distribution in the reflection level of the residual amorphous marks, which yield higher media noise.

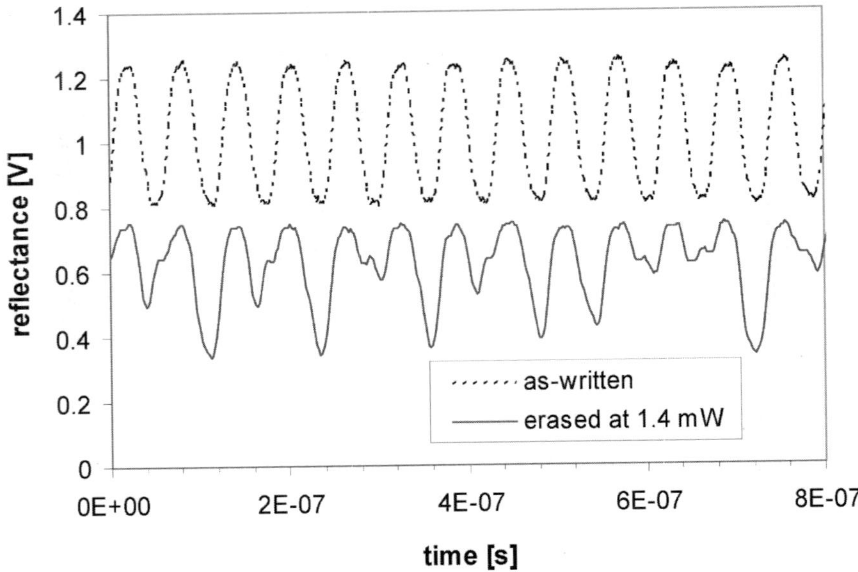

Figure 66. Upper-plot: time-resolved reflectivity of a track with an 8T single tone carrier (as written); lower plot: the resulting waveform after erase with 1.4 mW laser power.

Another cause for the appearance of the erase bump is as follows. In the recording stacks considered absorption of the amorphous state of the phase-change layer is higher than the absorption of the crystalline state. Therefore, for a given laser power amorphous marks experience a higher temperature rise than the crystalline spaces. This results in a selective melting process upon erasure within an erase power range close to the melt-threshold power of the crystalline state. During passage of the continuous laser beam with such erase power, the power is just high enough to induce melting of the amorphous marks while the material in the spaces remains crystalline.

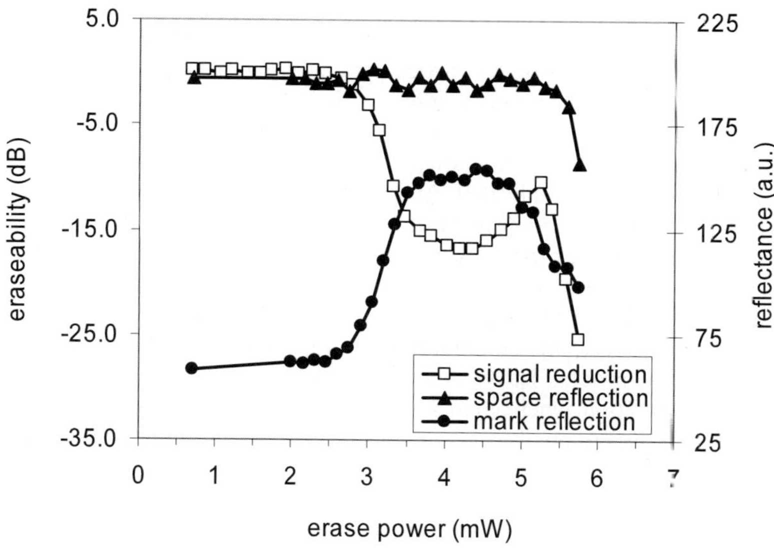

Figure 67. Erasability characteristics of an 8T data pattern: the CNR reduction and the reflection of the amorphous marks and crystalline spaces as a function of erase power.

The re-writing of marks by an erasing beam can also be derived from the reflectivity signals. In Figure 67 an erasability curve is plotted together with the reflectivity profiles. The experiment was done with a single-tone carrier where the mark/space length was larger than the spot size. In such a case the presence of spaces does not influence the reflectivity level of marks and vice versa. In this way reflectivity levels originating from marks and spaces can be detected separately. In the figure, full circles and full triangles denote the reflectivity levels corresponding to marks and spaces, respectively. The open squares represent the signal reduction upon erasure. If no significant erasure takes place the reflectivity levels and the signal generated by the data pattern remain practically unchanged. This corresponds to the erase power range up to about 2.5 mW in the plots. When the erasure sets on, the effective mark area becomes smaller (or rather the mark becomes narrower). This gives rise to an increase in the reflectance level originating from the mark and, in turn, leads to a smaller modulation and, therefore, a signal reduction. In these measurements, the erasure sets on at about 2.5 mW and continues up to about 4.5 mW With further increase of the erase power up to about 5.3 mW, reflectivity of the marks decreases while the reflectivity of the spaces still remains unchanged. A decrease in mark reflection occurs if the effective mark area (or rather mark width in this case) becomes larger. The larger marks give rise to a higher signal modulation and, hence, a higher CNR, which is seen as a peak in the CNR reduction profile at about 5.3 mW erase power. With yet higher erase powers both mark and space reflectivity levels decrease simultaneously. At these high erase powers the whole track becomes molten. Due to the quenching process taking place in the stack after the laser spot passage a continuous amorphous trace remains in the track and gives rise to the total reflectivity drop. This is also visible in images obtained with a Transmission Electron Microscope. In Figure 68 TEM images of an erase process, including re-writing of marks by an erase pulse, are given along with the erase curve. In this experiment, a single-tone data pattern was written on the disc and subsequently erased with a continuous erase pulse of a certain power. The six images correspond to the six data points in the erase curve as indicated. At 0.95 mW erase power (image 1) hardly any erasure takes place and the mark remains basically as it was written. At 2.0 mW and 2.2 mW (images 2 and 3) partial erasure occurs predominantly in the middle of the track leaving an H-shaped mark remainder. At about 2.5 mW (image 4) complete erase is achieved. The TEM image exhibits a ridge-like contrast pattern formed by the crystallization process of the amorphous mark. In the 2.4-3.0 mW power range (image 5) the erase process leaves amorphous line-shaped effects in the middle of the track at the position where marks have been situated. Though may be poorly visible due to the image quality, amorphous effects are discontinued at the spaces between the original marks. At higher erase powers (image 6) a continuous amorphous line is created in the middle of the track by the erasing beam.

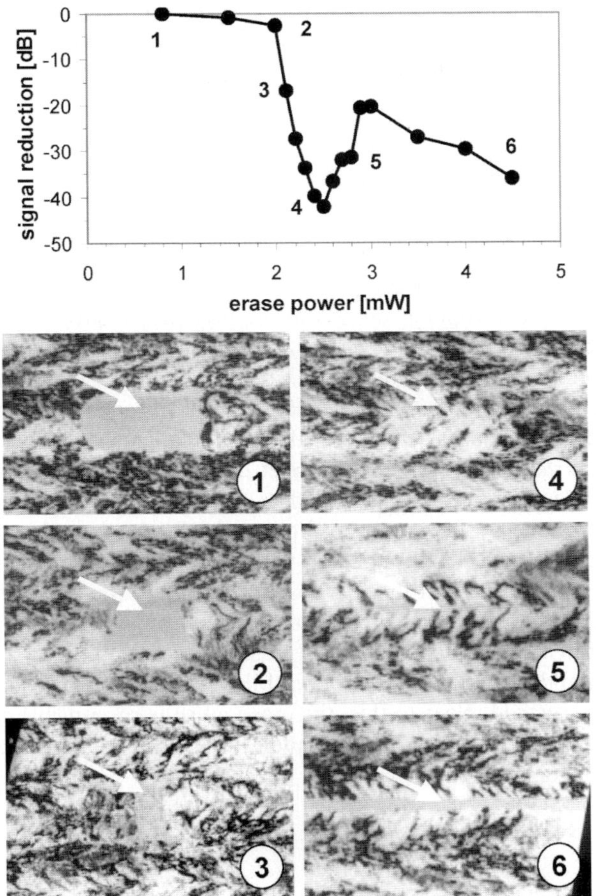

Figure 68. TEM images of an erase process at different erase powers. The corresponding erase curve with numbers is given at the top for guidance.

Designing a stack with equal absorption of the amorphous and crystalline state can reduce the erasability bump. In this way, the erase process can be improved in certain cases. However, taking into account all other requirements imposed on the recording stack such a solution cannot always be realized.

The erasability can further be improved by reducing the erase velocity. Results for different linear velocities of the moving laser spot are given in the lower panel in Figure 64. For the material considered complete re-crystallization at low erase powers is only achieved at the two lowest erase speeds shown in the figure, namely LV=4.2 and 5.25 m/s. Since the heat source moves relatively slow along the track, the temperature remains high for a longer time leading to complete re-crystallization. The temperature profiles are much narrower at higher linear velocities. Accordingly, the time for re-crystallization is then shorter. Furthermore, a shorter dwell time for higher erase velocities implies that more power is needed to attain the same temperature in the stack. Thus, for higher erase velocities more erase power is required in order to obtain the same mark width after re-crystallization.

4.1.4.   Write-erase sequence

As can be perceived from the erasability plots, for any phase-change material a combination of disc velocity and erase power level can be found that leads to sufficient erasure. However, good erasability is just one of the many requirements imposed on a rewritable optical disc. The requirement of direct overwrite implies that both old data are erased and new data are written in a single passage of the laser spot. A well-optimized erase element of a write strategy may at the same time result in bad write performance of this strategy causing, for instance, re-crystallization of a mark during writing or partial erasure of data in the adjacent tracks (a phenomenon that is called thermal cross-erase). If a direct overwrite approach is used, the write strategy should be designed for optimized erasure and recording of data. The disc velocity is determined by the data transfer rate that has to be achieved. Hence, the erasure has to be realized for this specific velocity. Furthermore, due to the fact that a recording stack has a certain thermal response to the write-pulse trains and erase pulses, writing and erasing processes influence each other when a direct overwrite principle is applied. Generally, an erase pulse is responsible for three aspects of recording. It should provide sufficient erasability, which is a measure of how well the old marks are erased. Next to this, the erase pulse is responsible for the longitudinal shape and the final length of the written marks. The latter two aspects will be discussed in more detail below.

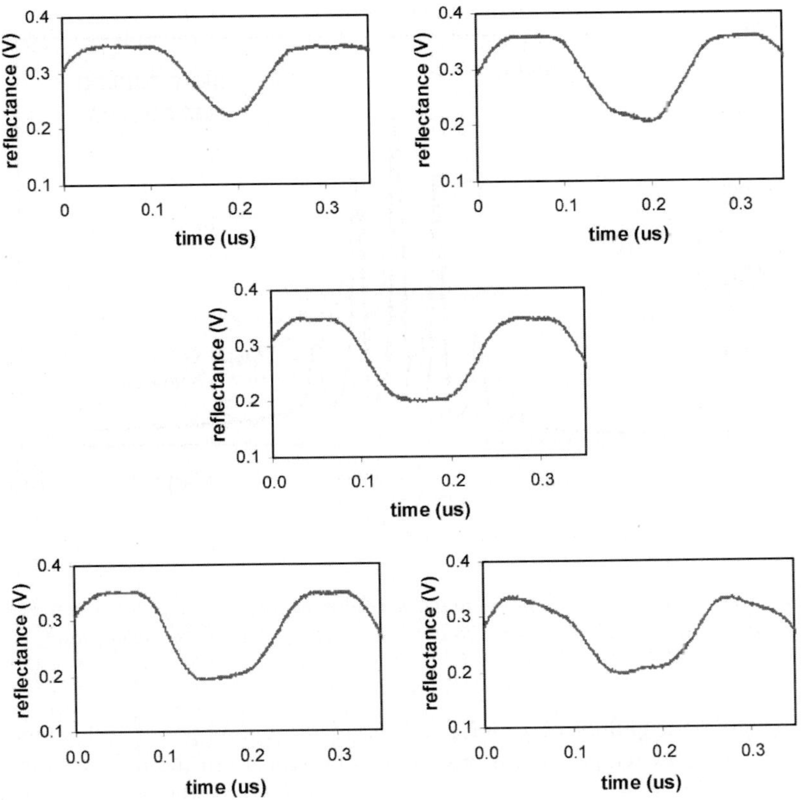

Figure 69. Time-resolved reflectivity of an 8T carrier written with the same write pulse sequence and write power but with different levels of the erase power, increasing from (a) to (e).

In Figure 69 a series of reflection modulation profiles is shown as measured from 8T BD single-tone carriers. The marks were written with a write strategy comprising pulse trains of 7 write pulses. Between the pulse-trains a continuous erase level was applied. The different reflection levels were obtained for different erase power levels, with the level increasing form panel (a) to panel (e). The pulse height of the write pulses in the pulse trains was constant for all cases. During DOW, the heat delivered by a write-pulse train adds up to the heat already present in the stack due to the preceding erase pulse. The stack's response to such a laser power profile results in a typical Gaussian shaped temporal temperature distribution profile. Examples of the temperature response to such a write strategy are given in Figure 70 for a slow-cooling and a fast-cooling recording stack.

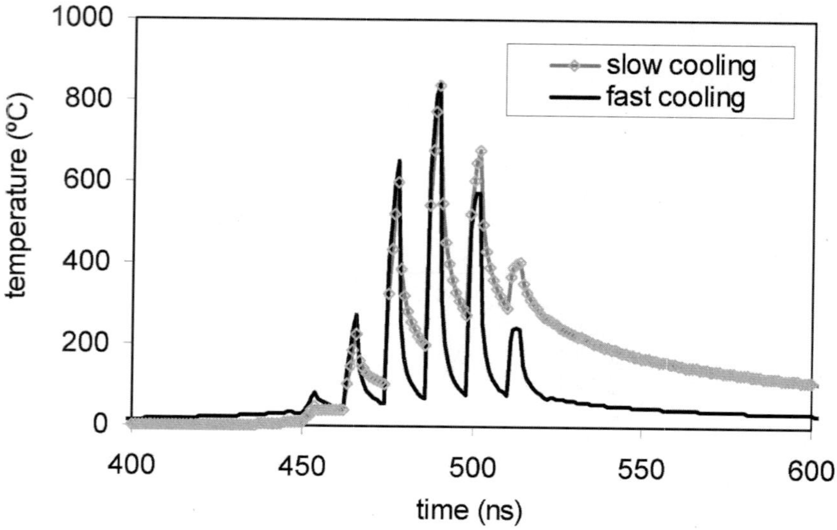

Figure 70. Temperature-time response to a laser pulse train of 6 write pulses superimposed to a DC erase level for a slow-cooling and a fast-cooling BD recording stack.

The applied write pulses of a pulse train lead to superimposed fast temperature elevations and fast temperature drops. As can be seen from the reflection traces in Figure 69 different erase power levels result in different reflection profiles of the marks written by the write-pulse trains. When the erase power is low, little heat is accumulated in the recording stack prior to an upcoming write-pulse train. As a result, the first pulses of the pulse-train heat the material to a temperature, which is just sufficient to melt a small area. As the write pulsing progresses more heat becomes accumulated and a broader area is molten. Such a thermal development in the stack leads to a mark with a narrow leading part and a broad trailing part. A mark with a narrow leading part yields higher reflection and lower modulation originating from this part of the mark (see panels (a) and (b)). If the erase power level of the write strategy is chosen adequately, a symmetric mark is obtained of equal width along the entire mark length (see panel (c)). If the erase-power level is high, the severe pre-heat caused by the erase level leads to a very broad area molten by the leading pulses of the following pulse train. This, in turn, results in a mark with a broad leading part and a narrow trailing part (see panels (d) and (e)).

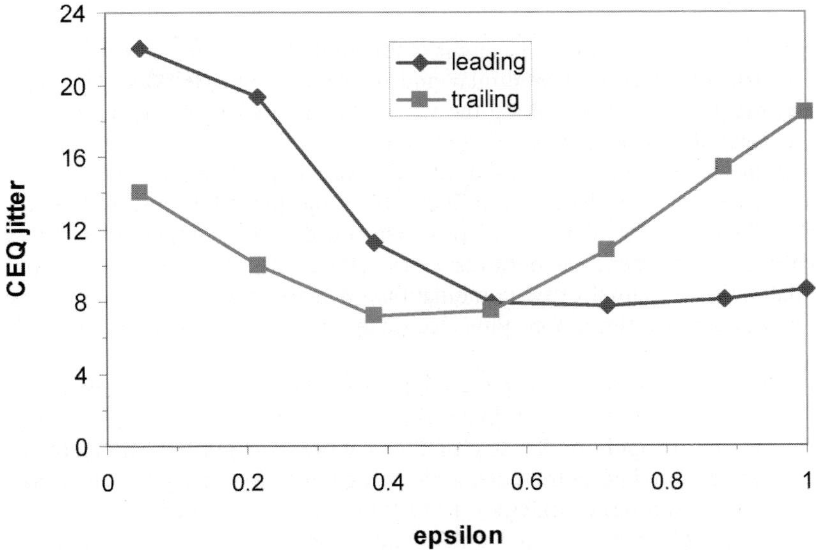

Figure 71. Conventional equalizer jitter of the trailing and leading edge of a 17PP random data pattern (23.3GB, TP=320 nm, single track).

A variation in the mark shape from its leading to its trailing part leads to a failure in the time-interval detection of the system. In Figure 71, leading and trailing jitters of a random data pattern are plotted as a function of the erase power level of the write strategy. The erase power is indicated here in terms of epsilon, which is the power ratio between the write pulse and the erase pulse in the write strategy. The random data pattern contained 2T to 8T marks and spaces randomly distributed. The data pattern was written in an empty track, the adjacent tracks were left empty to exclude the effect of optical cross-talk. The recording was done with the same write power, but with variable erase power. As can be seen from the figure, the leading edge of the marks dominates the poor performance at lower erase powers while the trailing edge of the marks spoils the performance at the high erase-power end. This observation is in line with the mark shape analysis. Low erase power leads to a mark with a narrow leading part and a nominal trailing width. Detection of the leading edge is, therefore, very sensitive to all kinds of noise contributions, which in turn is responsible for the steep increase in jitter at the leading edge. A similar argument holds for the situation with excessive erase power for which the trailing part of the mark is narrow. Detection of the trailing edge is then much more sensitive, which in turn is expressed in higher trailing jitter.

The erase-power level, which gives rise to a mark of nominal width along the entire mark stretch, may in general not be good enough to achieve sufficient erasability. To satisfy the requirement of both sufficient erasability and optimum writing performance, the write strategy is usually optimized together with the phase-change material's crystallization speed and the recording stack design. Ultimately, complete erasure should be achieved at the desired disc velocity and erase-power level.

4.1.5.   Write strategy types

From the optical drive point of view it is favorable that writing of each mark length from a data set is done in a similar and, preferably, simple way. The fewer and simpler the parameters describing the write strategy are, the lower the requirements on the laser driver would be. Furthermore, a less sophisticated optimum power calibration procedure could be used. In a first order approach, it can be stated that a given mark (space) is thermally influenced by the preceding and following space (mark). These effects of pre- and post heat need to be compensated for and are typically accounted for in an optimized write strategy. In the case of 17PP run length modulation code, a total of 64 combinations of marks and spaces are possible. For each combination, pulse power, pulse length and pulse position can be defined.

From detailed analysis of the phase-change recording process, it appears that in typical phase-change stacks within a wide disc velocity range each mark length can be well written by applying the so-called N-1 write strategy. In such a strategy, N-1 write pulses are applied in the pulse train to write an N-channel-bit long amorphous mark. Using such a write strategy for DVD implies that 3T marks would be written with 2 pulses, 4T marks - with 3 pulses and so on. One of the nice features of this write strategy is that increasing mark length by one channel bit is done by increasing the write-pulse train by one pulse. This one-by-one approach gives versatility in optimizing write strategy parameters. For higher disc velocities and for recording in slow-cooling semi-transparent stacks other types of write strategies are more suitable. In these cases cooling provided by the stack is usually not sufficient to suppress severe re-crystallization during writing of data. The solution to this problem has to be found in an improved write strategy. The use of the N-1 write strategy with significantly reduced pulse-train duties becomes inefficient due to write-pulse length limitations imposed by the laser and the laser driver. Also, such a write strategy with extremely short pulses requires a high laser power, which is typically not available in solid-state lasers. A proper solution to prevent re-crystallization during writing in high-speed applications and slow-cooling dual-layer stacks is reduction of the number of write pulses in the pulse train. Among a number of strategies with fewer pulses the 2T (N/2) and 3T (N/3) write strategies have proven to be most suitable for implementation. These strategies will be discussed in Chapter 5.

4.2.   Quality of recorded data

4.2.1.   Material-limited run-length jitter

As has been mentioned in Chapter 1, the edge-detection method is widely adopted in optical recording for retrieving data from the disc. Currently, much effort is put in advanced bit detection schemes that are based on maximum likelihood detection.

These detection schemes, in which the most-likely waveform is determined from the largest similarity between measured and calculated time-resolved waveforms, have the potential to increase the data density even further (either by channel bit length or track-pitch) or to increase the detection margins at the targeted density.

Edge detection imposes stringent requirements on the precision of spatial positioning of the mark edges and, therefore, demands subtle thermal balancing. If the write strategy and the recording medium are appropriately optimized, the position of the edges can be controlled within nanometer precision.

In an ideal medium characterized by perfect homogeneity and instant quenching the profile of the mark can be very smooth and would be determined by the melt-edge. However, in real discs quenching is not instantaneous and thermal properties on the micro-scale are influenced by the materials' morphology and microstructure. Furthermore, due to a relatively high thermal conductivity of actual recording stacks, writing of a mark affects the shape of the marks in the neighboring tracks by causing their partial crystallization at the side edges (the so-called cross-erase effect). Thus, in real phase-change media the formation of mark edges is determined by a crystallization process. As it is very well visualized by computer simulations (see Figure 48), an erase pulse following a pulse-train causes partial re-crystallization of the amorphous mark at the trailing edge giving it a crescent shape. In a granular (polycrystalline) material the crystallization speed within a crystalline grain differs from the crystallization speed at the grain boundaries. Depending on whether the trailing edge of a written mark falls within a single grain or in the vicinity of or at a grain boundary the amount of re-crystallization caused by the upcoming erase pulse is different. This difference causes deviations in lengths of otherwise equivalent marks. And the lower the cooling rate, the longer the re-crystallization period, and the larger the difference in the amount of re-crystallization. The influence of crystallites on the mark length is most pronounced in slowly cooling stacks based on FGM-type materials exhibiting relatively large crystalline grains. Figure 72 illustrates this phenomenon, where two two-dimensional inter-symbol-interference (ISI) plots are shown. The plots display time-interval distributions of different marks and spaces of an 2T-8T random-data pattern. Due to the constant disc velocity used during readout the time-interval distribution is proportional to the length distribution of the marks and spaces, which leads to a matrix of equidistantly spaced "clouds" of measured data points. The horizontal axes correspond to spaces; the vertical axes correspond to marks. Crossing points of the dashed grids indicate ideal lengths that are expected for a certain mark followed by a certain space. The tilt of the distribution "clouds" (and its orientation) reveals the obvious effect that the longer the mark is the shorter the next-following space becomes, and vice versa. The plots are measured on two BD media based on the same FGM phase-change alloy. The plot shown in panel (a) is acquired for a fast-cooling single-layer recording stack, which comprises a 120 nm thick Ag-alloy heat-sink layer. The plot in panel (b) is taken for a slow-cooling semi-transparent recording stack used in dual-layer media (see Chapter 5). Such a stack has a 10 nm thick Ag-alloy heat sink layer, which apparently leads to a much slower cooling behavior. Compact clouds that are perfectly centered on the dashed-grid crossings characterize the inter-symbol-interference plot of the fast-cooling stack. The compactness of clouds implies that the spread in mark and space length distributions (and, therefore, the jitter level) is rather low. Furthermore, the average mark and space length coincide with the ideal position indicated by the crossings in the dashed grid of the plot. The ISI plot for the slow-cooling stack has also well-centered clouds, but the clouds are much more spread out hinting to a very large distribution in the length of marks and spaces in this slow-cooling recording stack. Despite the fact that the write strategy used for the first disc is less well optimized this fast-cooling stack exhibits much smaller deviations in bit lengths and, consequently, a superior error rate.

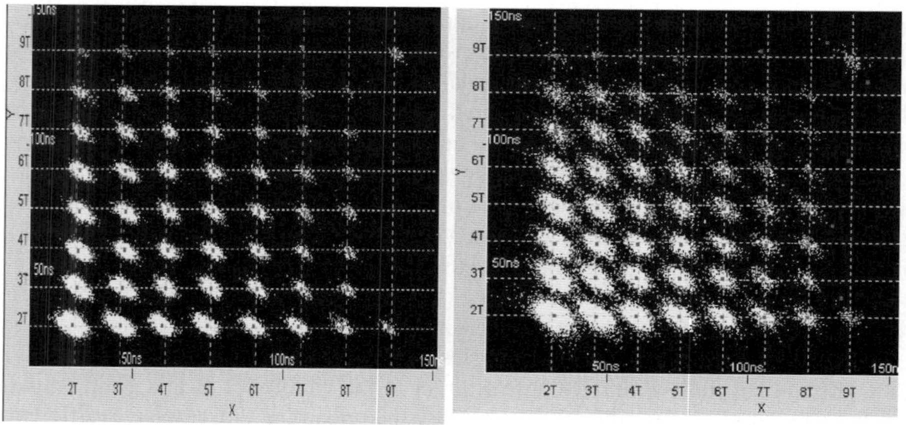

Figure 72. (a) Inter-symbol-interference plots of a 17PP data pattern (23.3 GB, BD conditions) written in (a) a fast-cooling stack and (b) a slow cooling stack.

4.2.2.   Overwrite stability

Next to the sources of media noise discussed earlier, recording stack deterioration caused by repetitive storage and retrieval of data adds to the overall media noise picture. These phenomena are often referred to as DOW-cyclability and repeated-read stability.

Figure 73. 8T marks written with 4 ns write pulse and 6 ns cooling gap as a function of the number of DOW cycles: track #1 DOW=1, track #2 DOW=10, track #3 DOW=100, track #4 DOW=1000. The recording velocity was LV=10.6m/s.

The number of direct overwrite cycles for a given recording medium is usually limited by degradation of the recording stack or deterioration of the written marks by severe re-crystallization during writing. The major reasons for this are mechanical damage of the stack caused by substantial and non-identical thermal expansion of the materials, by material flow, and by phase separation, which is fueled by out-diffusion and segregation of certain elements. The consequence of the latter is that media based on both NDM-type and GDM-type of phase-change materials exhibit an increase in crystallization rate (i.e. become faster) upon repetitive direct overwriting of data.

In Figure 73 a TEM image of four adjacent data tracks is given. Using direct-overwrite recording, an 8T carrier was written once in the first track, 10 times in the second track, 100 times in the third track and 1000 times in the fourth track, as indicated in the figure. The marks were recorded at 10.6 m/s at BD conditions. Two main observations can be made as the number of write cycles increases. First, the morphology of the crystalline matrix of the material changes significantly with the number of DOW cycles. Second, deterioration and size reduction of the amorphous marks take place with the increasing DOW cycles. The marks exhibit constriction phenomena at their center, which indicate that the material has become faster with the number of direct overwrite cycles.

The next experiment also illustrates that the phase-change material becomes faster with increasing number of direct overwrite cycles. A pattern of random data was written twice (DOW=2) in a fast-cooling BD phase-change stack at a recording velocity of LV=10.6 m/s. The TEM picture of this disc is given in

Figure 74. The marks exhibit long tails, which are believed to be caused by incomplete erasure of old data. Apparently, the velocity of crystal growth of the phase-change layer is not sufficiently high to enable full erasure at this high recording speed. The high recording velocity, which leads to a reduced dwell time for re-crystallization of the old marks, also contributes to the incomplete erasure as well.

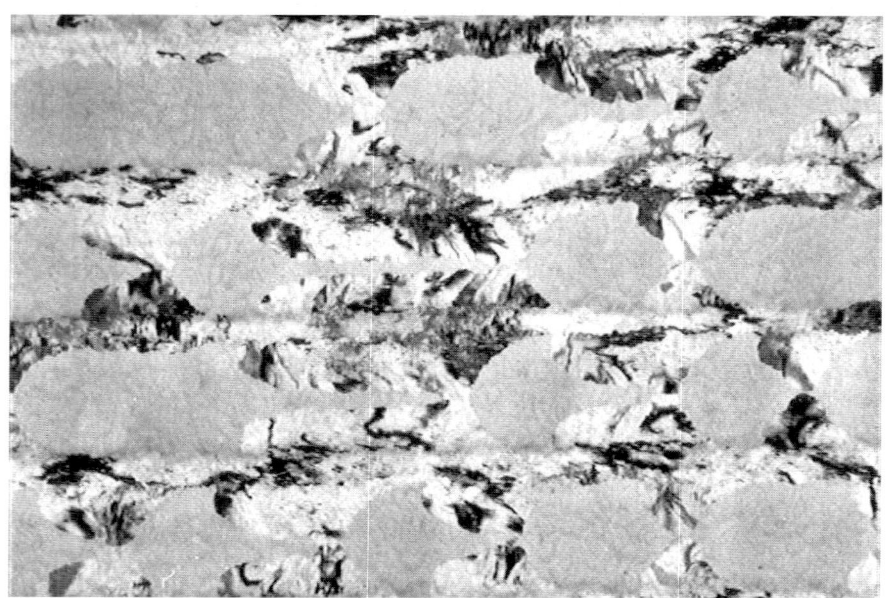

Figure 74. Random data for DOW=2 at LV=10.6 m/s (MIPI stack).

Results for this disc recorded at high linear velocities for DOW=1000 are given in Figure 75. The amorphous tracks in between the amorphous marks have disappeared. The high number of direct overwrite cycles made the phase-change material faster. The erase level, and the effect of therewith-created temperature-time history, appears to be sufficient for complete erasure of the old marks. Also the width of the marks, and thus the resulting modulation, is significantly reduced at this higher recording speed. A possible explanation is that the material has become faster due to the large number of direct overwrite cycles. Re-crystallization during writing results in a narrower mark. Also thermal cross-write may lead to partial re-crystallization at the mark sides.

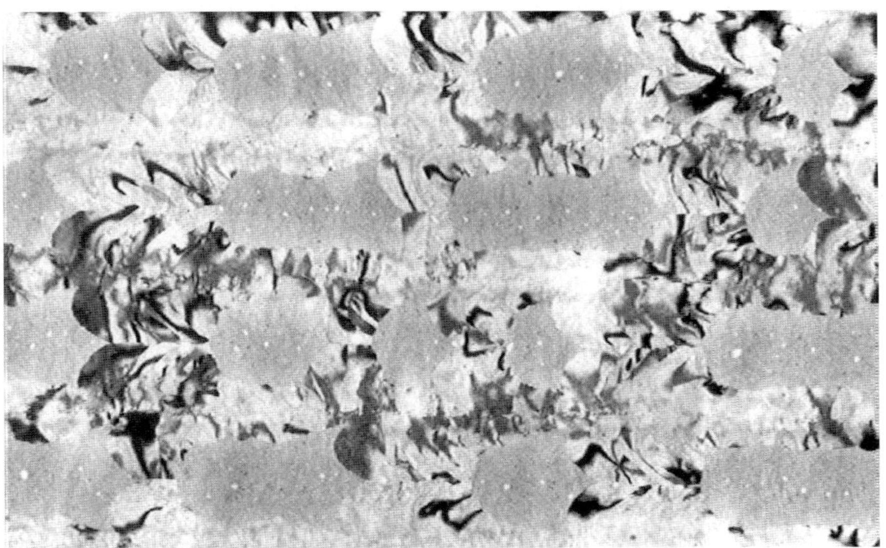

Figure 75. Random data for DOW=1000 at LV=10.6 m/s (MIPI stack).

Another remarkable observation is the presence of the small holes inside the marks at DOW=1000, uncovering mark deterioration. These small holes are most probably caused by material migration through the phase change layer due to repetitive thermal expansion and compression.

The deterioration caused by DOW is an irreversible process and is accompanied with rising media noise, a decrease in signal modulation, and an increase in variation of the mark edge positions on the disc. As a consequence, a steep increase in jitter and SER is observed. This is demonstrated in Figure 76, where jitter is plotted versus the number of DOW-cycles.

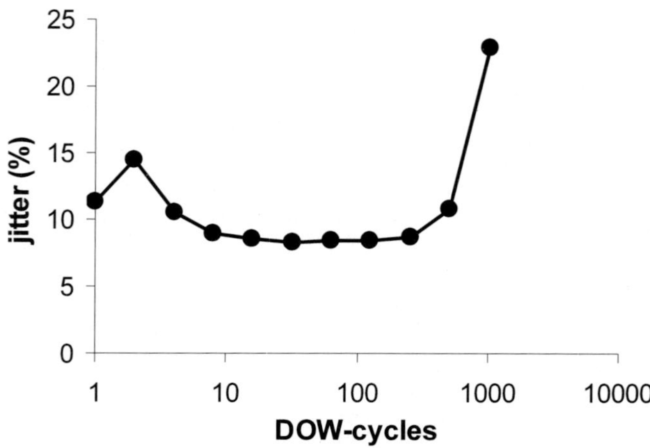

Figure 76. Jitters as a function of the number of direct overwrite cycles.

One more feature of the overwrite performance of phase-change media is a jitter bump observed during the first several overwrite cycles (typically DOW<10). The origin of the bump lies in the difference in the properties of the phase-change layer before and after recording. The atomic arrangement of the as-deposited amorphous state is different from that of the amorphous state formed from the well-ordered crystalline state by melt quenching. Furthermore, the crystalline state formed from the as-deposited amorphous state by initialization exhibits poor long-range order in the atomic arrangement, high amount of defects, grain boundaries, and strain. During repetitive overwriting high temperatures induce high atomic mobility, which promotes atomic rearrangement and finally leads to a settled-down state. This transition stadium is typically overcome within several overwrite times. In principle, write strategies can be optimized for each state of the phase-change material's transition stadium. However, since in a rewritable system the number of times the data are overwritten is large and, thus, a vast majority of recording will take place in the settled-down state of the phase-change material write strategies are usually optimized for DOW=10. Generally speaking, such a write strategy is not optimal for the first a few overwrite cycles and, therefore, may yield worse performance and higher jitters.

4.2.3.   *Repeated read stability*

From the detection point of view it is always favorable that the noise level of the system is kept to a minimum. In CD and DVD systems SNR is predominantly media noise limited. For high-density storage systems, such as BD, the issue of other noise sources appears to be increasingly important since electronics and laser noise become limiting at the high-frequency end of the data spectrum. One of the noise sources in an optical storage system is the laser noise. Lasers are typically character-ized with a relative intensity noise (RIN) level, which is lower for higher laser powers. From this perspective it is beneficial to use higher laser powers during read-out. However, in application to phase-change media repetitive read-out can result in data erasure even when the read powers are far below the erase power level. By contrast to DOW-cyclability such data deterioration is reversible in the sense that data can in principle be re-written on the same place on the disc without any appre-ciable loss of quality. In Figure 77, the symbol error rate (SER) is plotted versus the number of read cycles for two different read power levels. (The erase power level for this medium is 3.3 mW). As can be seen from the figure, after a certain number of read cycles at 0.8 mW read power a steep increase in error rate takes place. Furthermore, as can be learned from the experiments the higher the read power the fewer the number of cycles that cause deterioration.

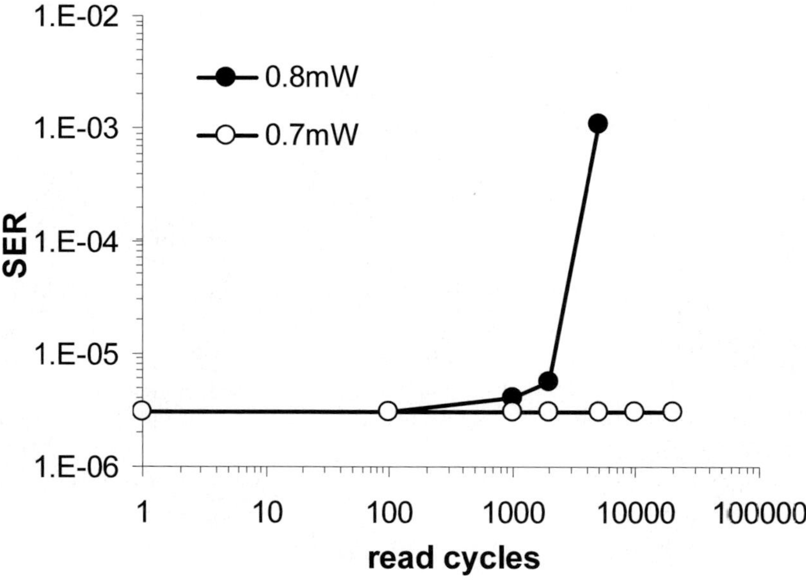

Figure 77. Symbol error rate versus the number of read-out passages of the laser spot over the same data track as measured at read powers of 0.7 and 0.8 mW.

TEM analysis provides insight into the physical mechanism behind the deterioration of the amorphous marks. For this analysis, a pattern of random data was written (DOW=10) on a GDM-base BD medium and repetitively read out at a disc velocity of LV=5.28 m/s and a read power of 0.65 mW. Measurements were done for various numbers of read cycles, between 512 (2^9) and 131072 (2^{17}). TEM images of marks subjected to 2^{17} read cycles are given in Figure 78. As has been discussed in the preceding chapters, GDM-type of materials exhibit a very low nucleation probability and crystallization of amorphous marks usually takes place from the outer mark edges inwards.

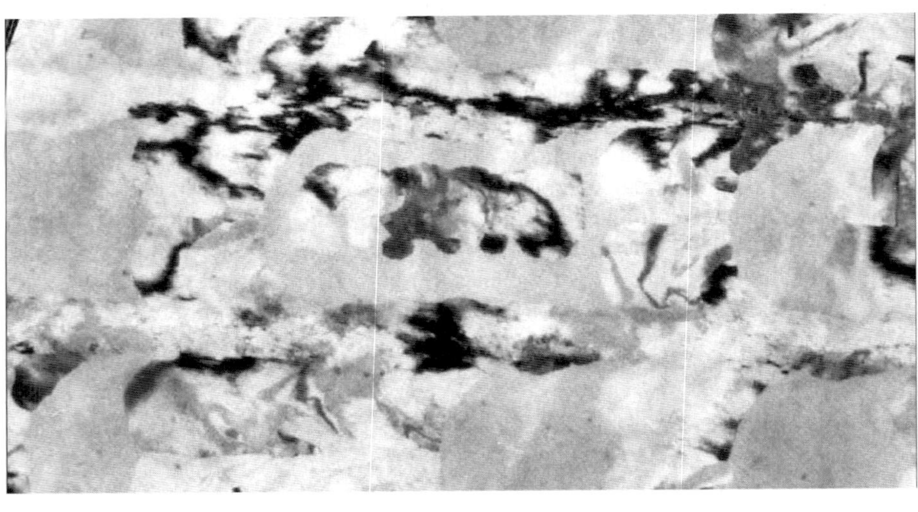

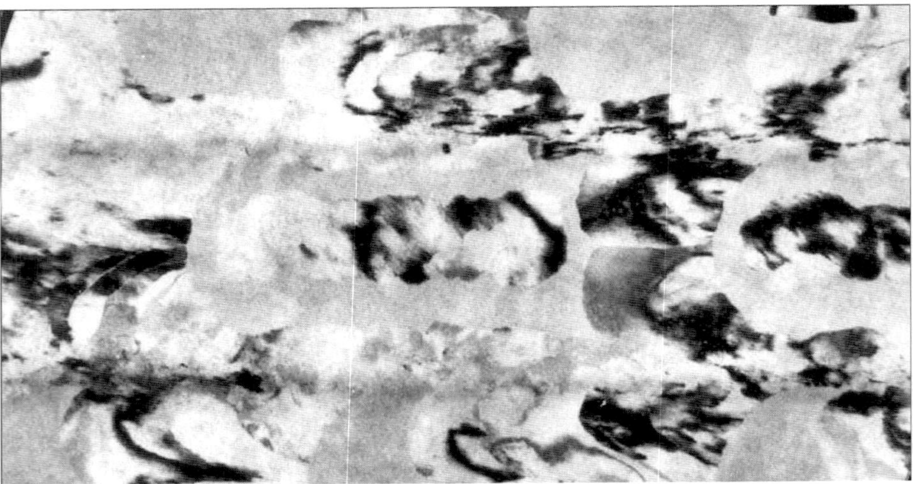

Figure 78. TEM pictures of amorphous marks (DOW=10) after 131072 (2^{17}) read cycles (read power P_r=0.65 mW).

Therefore, re-crystallization upon repeated read can be expected to occur via crystal growth from the crystalline-amorphous interface at the mark sides. In particular, significant re-crystallization at the leading and trailing part of the mark could be expected because of the resulting temperature distribution during repeated read. The temperature distribution induced by the focused laser has in first order approximation a Gaussian shape in the cross-track direction. The temperature experienced at the leading and trailing part of the mark is, therefore, much higher than at the mark sides. Peculiarly, TEM images reveal that upon repeated read re-crystallization takes place from the inner mark area. What is also interesting is that the remaining amorphous rim has more or less the same width, independent of the original mark length (3T to 8T). The most plausible explanation for these observations is the absorption difference between the amorphous mark and the crystalline phase (background, spaces) in combination with the nucleation component of the crystallization process. Although GDM-type materials have a low nucleation probability, it is not necessarily zero. A stable nucleus may be formed for sufficient read cycles, from which crystal growth may occur. When the laser spot passes the crystalline space between two amorphous marks, the temperature rise is not sufficient to induce crystallization from the leading edge. When the spot encounters an amorphous mark, more laser light is absorbed at the interior of the mark. During passage over the amorphous mark the amount of absorbed light and, thus, the resulting temperature increases. This mechanism would cause marks to re-crystallize from the inner side towards the edges. Since a maximum temperature is expected at the tail of the mark, re-crystallization is expected to start from the tail. Recorder experiments indeed show that the mark degradation starts at the trailing part of the mark. Furthermore, it was observed that longer marks deteriorate first, which is also explained from the higher temperature rise in case more amorphous material is encountered during the laser passage. The reason that erasure stops at a reproducible distance away from the edge can be explained by the same absorption differences. With a higher crystalline fraction in the (partly) amorphous mark, the absorption drops accordingly, which leads in turn to lower temperatures. This temperature reduction decelerates crystal growth to eventually end up with a narrow amorphous rim. The temperature rise is simply not sufficient anymore to sustain further crystallization.

The TEM pictures suggest a polycrystalline inner region of the deteriorated marks as can be judged from the optical contrast of the image. However, tilting the TEM sample and, thus, studying the behavior of the electron diffraction contrast over the crystalline area reveals that in a vast majority of cases the inner region is predominantly mono-crystalline. This observation suggests that only one nucleus was responsible for the re-crystallization. Keeping in mind the type of phase-change material used in the experiment, this is not surprising. Once a crystalline nucleus is formed, it quickly grows out in all directions of sufficiently high temperature. This sudden and quick disappearance of marks during read-out can also be well seen on an oscilloscope by tracing deterioration of signal modulation. It was observed that a mark could survive a large number of read cycles but once re-crystallization induced by repeated read set off it was completed within a few read cycles.

From TEM analysis, the number of partly deteriorated marks was determined out of a total of 100 marks for each number of read cycles. This deteriorated fraction is plotted versus the number of read cycles in Figure 79. For a few number of read cycles, no deteriorated marks were observed, but the number increased rapidly with increasing number of read cycles.

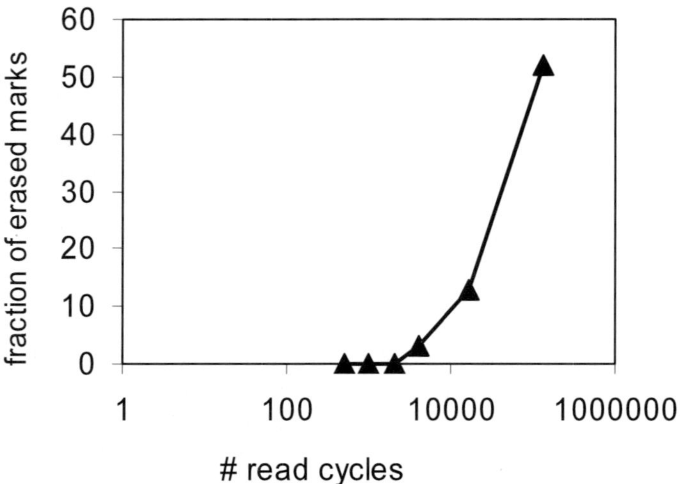

Figure 79. Fraction of deteriorated marks (DOW=10) as a function of read cycles (read power P_r=0.65 mW).

4.3. Effect of multi-track recording

To fit a maximum amount of marks on the disc surface, the data tracks should be placed as close to each other as possible. However, the proximity of data tracks has a direct influence on the quality of the data recording and retrieval process due to the finite size of the optical and thermal spot generated by the focused laser beam on the disc. Below, two major consequences of data track proximity are addressed, namely thermal cross-write and optical cross talk.

4.3.1. Thermal cross-write

Thermal cross-write is the phenomenon that writing data in a track causes partial erasure of amorphous marks in the adjacent tracks. If the thermal spot is much broader than the track pitch, the temperature elevation in the neighboring tracks is such that re-crystallization at the mark sides may occur leading to a reduced mark width and eventually to loss of data. A cross-write or multi-track experiment is typically performed by writing a data pattern in the central track and subsequently writing data patterns (possibly with overpower) in the adjacent tracks.

To illustrate the effect of cross-write, a TEM image is given in Figure 80. For this experiment, long 8T marks (BD-system) were written in the central track (DOW=10) and subsequently short 2T carriers for one overwrite cycle (DOW=1) were written in the adjacent tracks. The recording was done at LV=5.28 m/s disc speed. A clear constriction occurs at the location were the marks in the adjacent tracks appear.

Figure 80. Cross-write experiment of 8T marks written in the central track (DOW=10). Subsequently, short 2T marks were written once (DOW=1) in the adjacent tracks.

Cross-write behavior of short 2T marks (DOW=10) is illustrated in Figure 81 for long 8T marks written (DOW=1) in the adjacent tracks. Re-crystallization occurs at places where the long marks coincide with the 2T marks.

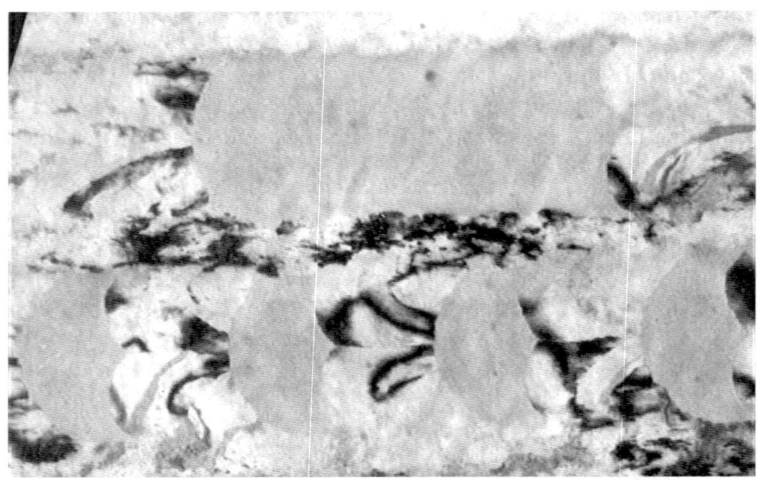

Figure 81. Cross-write experiment of 2T marks written in the central track (DOW=10). Subsequently, long 8T marks were written once (DOW=1) in the adjacent tracks.

The cross-write effect is one of the main factors limiting the radial storage density in phase- change media. As has been shown in Chapter 3, the effect of cross-write can be reduced by optimizing the track pitch and by tuning the groove-structure profile with respect to the laser power absorption.

4.3.2.   Optical cross-talk

Optical cross talk is the phenomenon when signal streams from several sources interfere with each other. This occurs e.g. if an appreciable part of the laser spot intensity covers adjacent tracks. In this case the same beam to the detector carries information, which originates from several tracks, simultaneously. The modulation profile of the central track is altered by the presence of marks and spaces in the adjacent tracks. Since the data in the neighboring tracks are non-correlated, such an effect causes additional noise in the detection channel.

An elegant way to solve the cross-talk issue is to apply a three-spot cross-talk cancellation scheme. In three-spot optics the main (central) spot is complemented with two satellite spots, which are placed at a fixed distance (in both radial and tangential direction) from the central spot. The satellite spots can, for instance, be created with a grating (see Chapter 1). The high-frequency information detected by the satellite spots is then processed to extract the single-track information that was detected by the central spot.

5. *Recording media*

Data transfer rate and data capacity are the appealing performance factors of optical recording systems. The main developments in optical phase-change media are, therefore, oriented towards materials systems that allow for a high recording speed and that enable dual-layer or even multi-layer recording. The first section of this chapter deals with dual-layer media, in particular meant for BD-RE recording. In the second section, phase-change materials for high-speed recoding are discussed. The emphasis is on DVD+RW and CD-RW high-speed recording.

5.1. *Dual-layer media and recording*

5.1.1. *Data capacity increase*

There are several ways to increase the data capacity of an optical disc. This can be done by increasing the storage density, by making use of multiple data layers, or by optimizing the format efficiency including the coding schemes, land-groove or groove-only recording, *etc*. The format issues fall out of the scope of this book and, therefore, will not be touched upon any further.

It has been explained in Chapter 1 that the storage density increase of an optical disc is achieved by scaling down the size of the recorded marks (or pits) representing data bits. In this way (i) shorter mark lengths lead to higher linear density along the data track and (ii) narrower mark widths allow to decrease the pitch of the data tracks on the disc and, thus, to gain in the traverse density. The size of the smallest effect that can be written is predominantly determined by the smallest size of the spot, which the laser beam can be focused to. The smallest spot diameter, in turn, is defined by the diffraction limit and can be expressed as $d=\lambda/(2NA)$, where λ is the laser wavelength and NA is the strength (numerical aperture) of the objective lens. Thus, higher recording densities are obtained by choosing lasers with shorter wavelength and objectives with higher NA. The result of such a two-dimensional data density increase is demonstrated in Table 1 of Chapter 1 for the three generations of optical media: CD, DVD, and BD. To further increase the data capacity, the third spatial dimension can be used. The simplest way is known from the old times of vinyl LPs: a double-sided disc. In this way the capacity of a disc can be doubled. The disadvantage of this way is the physical access of the data: the disc needs to be turned over or a relatively expensive double-head drive has to be used. A more elegant way to make use of the third dimension is by developing multi-layer media. The important feature of such media is that the same pick-up head from the same side of the disc can access all the data layers. The number of the data layers that can be put in one disc without severely affecting its playability is mostly limited by the optics, electronics, and mechanics of the optical drives. From the media design point of view, the major consequences that optical drive limitations impose onto the media design are the thickness of the transparent substrate, cover, and the spacer layers, as well as the reflectivity levels and transmittance of the data layers. For BD and future-generation systems the amount of disc warp becomes also critical.

5.1.2.   Layout of dual-layer media

In the case of rewritable media based on the currently available phase-change materials only two data layers can be allowed per disc side. The limitation is dictated by the optical properties of the materials and the drive electronics. A layout of dual-layer disc configuration is given in Figure 82. In the case of DVD, the two data layers are situated between two substrates and are separated by a spacer layer that is about 50 μm thick. The data layer that first faces the incident laser beam is semi-transparent at the wavelength applied and is called L0. The other data layer is called L1. In the case of BD, the two data layers are separated by an about 25 μm thick spacer, are placed on a 1.1 mm thick substrate and are covered with an about 75 μm thick cover layer. In this system, the indexing of the data layers is inversed. Namely, the semi-transparent data layer that first faces the incident laser beam is called L1 and the other data layer is called L0. In this chapter, the BD system will be mostly considered as an example to discuss the issue of dual-layer phase-change media. For this reason BD-indexing of the data layers will be used.

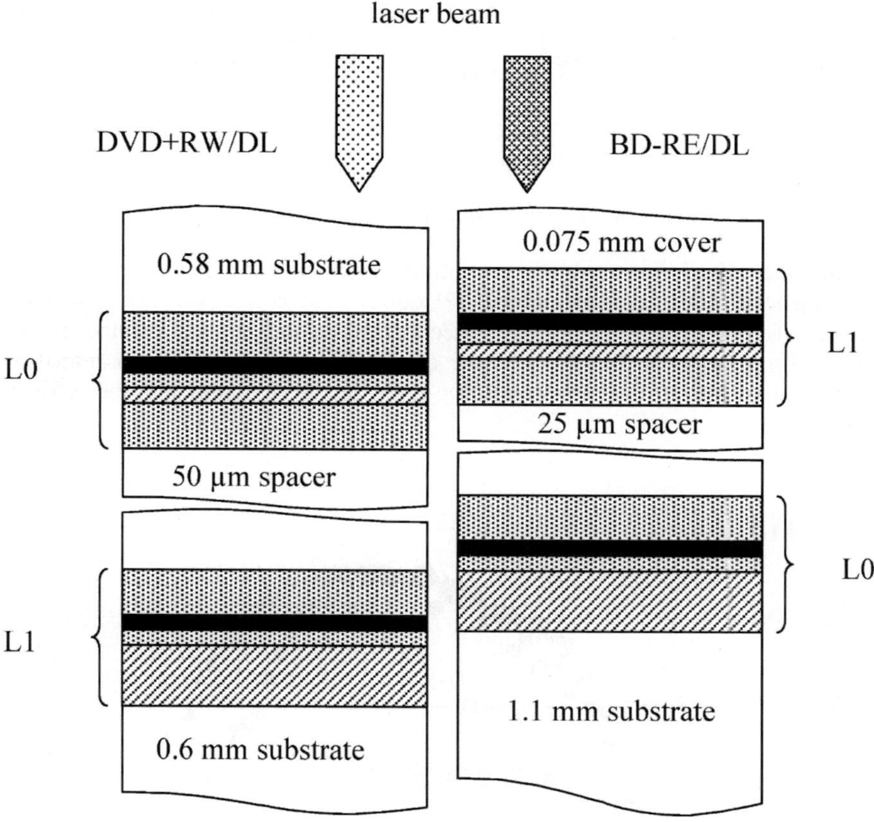

Figure 82. Schematic layout of DVD and BD dual-layer rewritable discs. The discs comprise two recording stacks, L0 and L1, one of which is semitransparent at the wavelength applied.

In the case of a dual-layer disc it is necessary that the L1 data layer has a sufficiently high transmittance at the laser wavelength applied in order to allow access of the deeper lying L0 data layer. For an optimum performance in a dual-layer configuration utilizing currently available materials it is favorable that the transmittance of the semitransparent data layer is about 50%. Typically, the transmittance of the phase-change recording stack used in a conventional single-layer rewritable disc is about nil. Such a recording stack comprises a number of layers two of which absorb the majority of the incident light. These are the metal layer and the phase-change layer. The metal layer is typically 120 nm to 150 nm thick. It usually consists of Ag- or Al-alloy and serves as a reflector and a heat sink (see Chapter 3). The phase-change layer has a thickness of typically 10-15 nm and accounts for some 30-40% absorption. To achieve high transmittance, which is necessary for the semitransparent L1, the thickness of these layers has to be reduced.

The layer thickness reduction inevitably affects all other important characteristics of the recording stack including reflectance, optical contrast, recording sensitivity and crystallization speed. All theses issues are dealt with in the sub-section below.

5.1.3.   Optical design of dual-layer media

For the sake of high transmittance, it is attractive to leave the metal layer out of the recording stack and consider a simple IPI design. The dielectric I-layers would be necessary for the reasons of thermal protection and optical contrast enhancement, as explained in Chapter 3. In Figure 83 the calculated maximum optical contrast of a number of phase-change materials is given as a function of the effective transmittance.

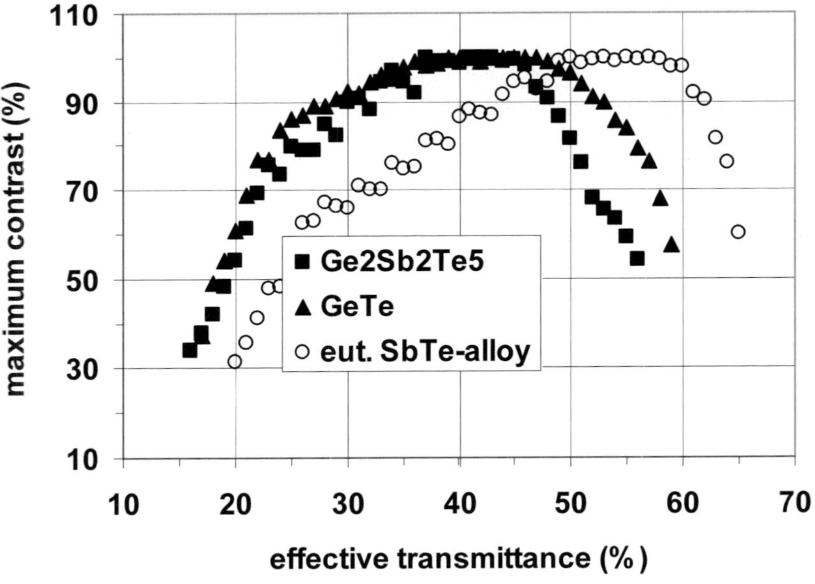

Figure 83. Calculated dependence of maximum optical contrast on effective transmittance of phase-change layers. See text for more details.

In the calculations, thicknesses of all three layers of the IPI-stacks were optimized. The dielectric layer thicknesses were varied in the $0-\lambda/2n$ nm range, where λ=405 nm is the laser wavelength and n is the refractive layer of the material. The thickness of the phase-change layers was varied in the 4–12 nm range. The effective transmittance is defined here as three quarters of the transmittance of the crystalline state and one quarter of the transmittance of the amorphous state. This is approximately the ratio between the areas covered with amorphous marks and the crystalline background, which fall into the laser beam passing through a fully recorded data layer (groove-only format is assumed). The optical contrast between the crystalline and amorphous state of the material is defined as the difference between the reflectance of the crystalline and amorphous state - reflectance of the crystalline state minus reflectance of the amorphous state - divided by the maximum of the two-reflectance levels. In this case, negative contrast values correspond to the optical stack designs where crystalline reflectance is lower than the amorphous one. Such stacks would result in low-to-high polarity of the high-frequency data signal. Since, on the one hand, optical parameters of the stack are mutually dependent and, on the other hand, the reflectivity level of the medium - in particular, in its initial state - should be high enough to make focusing and tracking possible, a boundary condition for the crystalline reflectivity has to be imposed. By experience, it is sensible to consider in recording stack designs where the reflectivity levels of the initial (*i.e.* crystalline) state of which is higher than 4%. This boundary condition is mostly dictated by the optical efficiency of the light-path and the electronics of the drives. Furthermore, optical contrast of the recording stack should preferably be above 80% in order to provide sufficient signal modulation. Keeping these requirements in mind, a number of useful facts can be learnt from the calculations. Firstly, materials at the GeTe- side of the GeTe-Sb_2Te_3 tie line and doped Sb_2Te alloys are the better candidates for use in the L1 data layer, as far as the optics is concerned. Secondly, to provide sufficient transmittance the thickness of the phase-change layers based on these materials has to be in the 5 to 10 nm range. Besides the optical characteristics, recording performance imposes a severe limitation on the smallest phase-change layer thickness. Generally, the vast majority of phase-change materials with layer thickness below about 5-6 nm has poor homogeneity leading to a dramatic decrease in the number of direct overwrite cycles. Furthermore, interfaces with the adjacent dielectric layers start to play a dominant role affecting the crystallization behavior of the phase-change material.

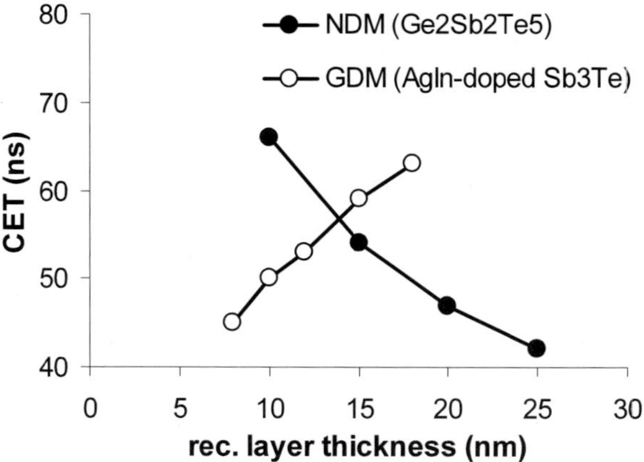

Figure 84. Dependence of complete erasure time (CET) on the layer thickness for NDM and GDM type of materials.

The effect of interfaces is demonstrated in Figure 84, where the time necessary to crystallize an amorphous mark is plotted versus the layer thickness for different types of phase-change materials. In the case of AgInSbTe GDM-material, reducing the layer thickness from 25 nm down to 8 nm results in shorter crystallization times. This means that the thinner the phase-change layer is the higher the maximum data erase speed and, ultimately, the higher the data transfer rate can be achieved. Below about 7 nm a steep increase in crystallization time is observed, see Figure 84 In the case of $Ge_2Sb_2Te_5$ NDM-materials, phase-change layer thickness reduction leads to a monotonous increase in the crystallization time and, thus, to lower maximum data transfer rates. Although not shown in the figure, for thicknesses below some 7 nm the achievable crystallization speed and, consequently, the data transfer rate become very low for both NDM and GDM materials.

In the remainder of section 5.1 semi-transparent stacks based on GDM type of materials will be considered.

5.1.4. *Mark formation*

Let's now look at the recording performance of a semitransparent layer that has a simple I_2PI_1 design. The P-layer considered here is a 6 nm thick GeInSbTe alloy. The I-layers are $(ZnS)_{80}:(SiO_2)_{20}$.

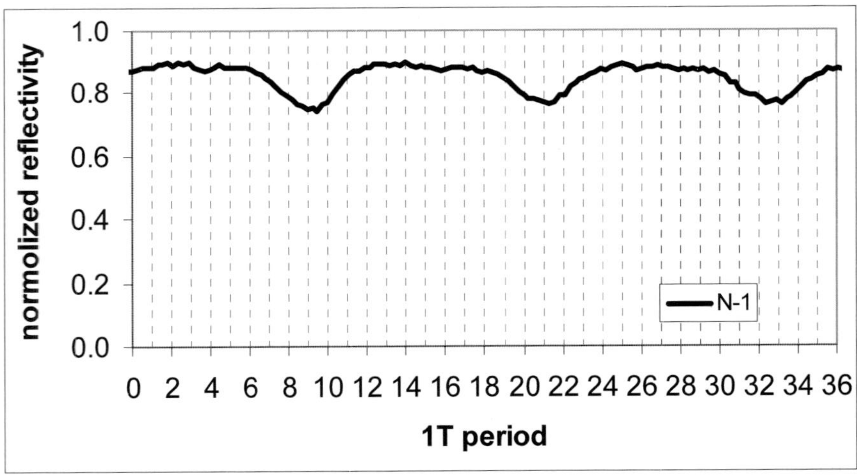

Figure 85. High-frequency signal of I6 carriers written in an IPI-type stack using N-1 write strategy.

Writing an 6T data carrier in such a recording layer using conventional N-1 write strategy results in the high-frequency signal pattern shown in Figure 85 with a solid line. Two striking peculiarities can be observed in the signal profile. First, the signal modulation is just 16%, which is much lower than can be expected on the basis of the 90% optical contrast of the stack. Second, the recorded marks are much too short and the spaces between them are much too long. The clue for these peculiarities can be found by performing thermal modeling of the recording process and TEM-measurements of recorded data. In Figure 86 the temperature response to a pulse train of 10 write pulses is shown for an I_2PI_1-stack (black solid line). For comparison, a temperature profile calculated for a conventional single-layer MIPI stack is shown as a grey line with box markers in the same figure.

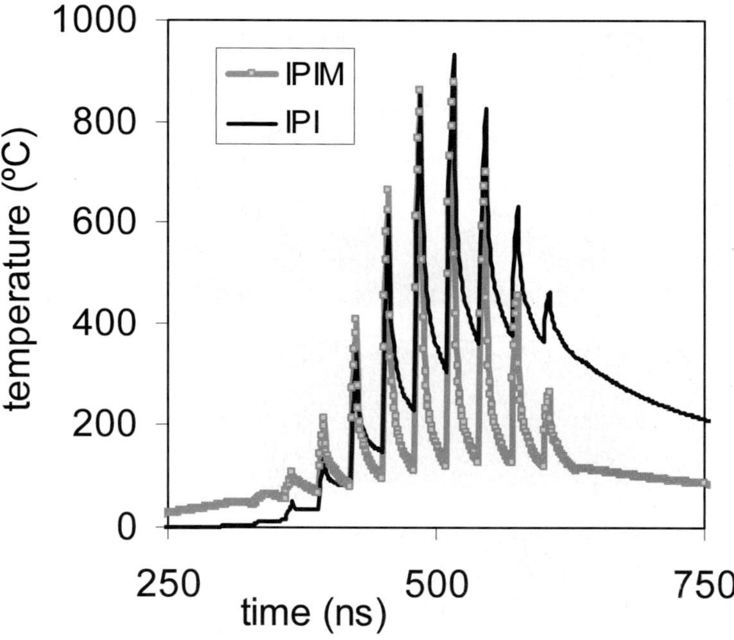

Figure 86. Temperature response to a pulse train of 10 write pulses, monitored in the center of the recording layer of an IPI and IPIM recording stack.

Each spike in the temperature profile corresponds to a single write pulse in the pulse train. The bump-like base of the profile reflects the total heat accumulated in the stack during writing. It appears that due to the absence of a heat-sink layer in the IPI stack too much heat is accumulated during writing in this stack compared to the conventional MIPI-stack. The excessive heat leads to severe re-crystallization of the amorphous mark being written. Detailed analysis of the simulations reveals that two processes contribute to this: (i) re-crystallization of the molten material during the cooling-down periods in the pulse-train; (ii) re-crystallization of the previously formed amorphous dot due to the heat induced by the subsequent write pulses in the pulse-train and due to the total heat accumulated in the stack. The latter of the two processes gives the main contribution to the reduction of the mark size. An example of the mark shapes obtained by modeling for IPI stack and MIPI stack is displayed correspondingly in panels (a) and (c) of Figure 87.

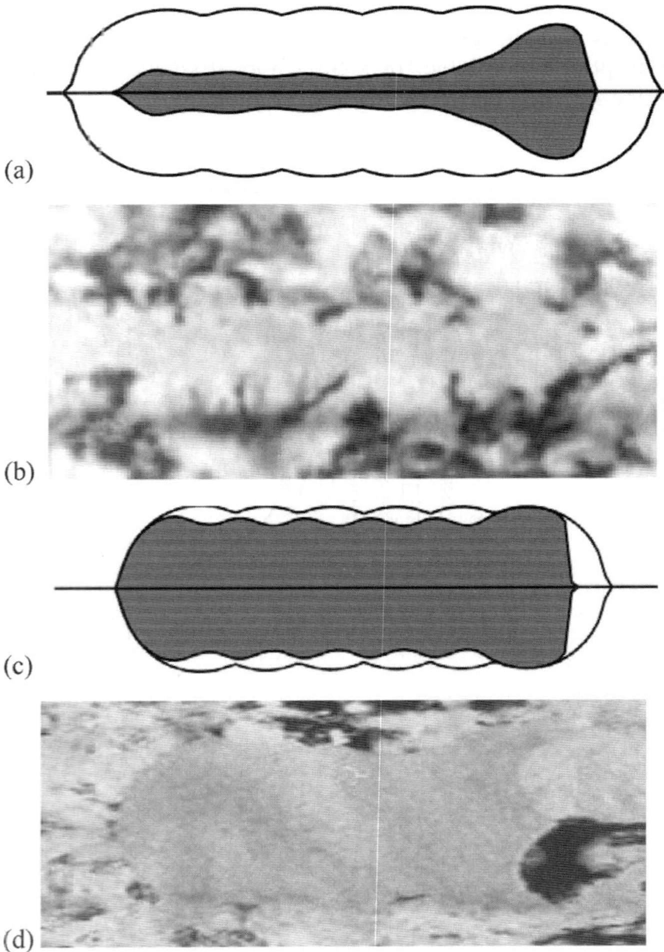

Figure 87. Typical mark shapes in an IPI (a, b) and an IPIM (c, d) recording stack as a result of N-1 write strategy. The upper row represents the results of computer simulations, the lower row denotes the results of transmission electron microscopy (TEM) analysis of recorded marks.

The solid-line contour in these panels depicts the edge of the molten area. The shaded field represents the final mark size left after the recording process is complete. The area between the solid-line contour and the shaded field is the area of the phase-change layer that undergoes re-crystallization. As can be seen, in the case of IPI stack the mark has a narrow "body" and a pronounced "head". In panels (b) and (d) of the same figure TEM images of the real amorphous marks are presented. A close similarity between the real marks and the results of modeling can be seen. In the case of a conventional MI_2PI_1-stack the thick metal heat sink provides a high quenching rate in the recording stack. As a consequence, re-crystallization of the mark during the cool-down period is suppressed (see the difference between the molten area and the final mark side in panel (c) of Figure 87). By contrast, in the case of an I_2PI_1-stack severe re-crystallization occurs and a very narrow mark is formed. The broader trailing part of the mark is due to the fact that the last write pulse of the pulse train is followed by a low-power erase pulse: The amorphous dot formed by the last write pulse is re-crystallized just slightly, leading to a large "head" of the mark. As a result, the "body" of such a mark causes almost no signal modulation and only a very low modulation is obtained from the "head" of the mark (see Figure 87). Optimization of write power does not lead to improvement in the mark shape. If lower powers are used, a very small area of the phase-change material is molten and the final mark remains too narrow. If higher write powers are applied, a yet larger amount of heat is accumulated in the stack resulting in complete re-crystallization of the molten area with no amorphous mark left after recording. From the analysis above it becomes obvious that the N-1 write strategy is not suitable for recording in IPI type of stacks.

To improve recording in an IPI stack or, more specifically, to be able to create marks of sufficient length and widths a few potential solutions can be considered. These solutions are discussed in detail in the upcoming subsections.

5.1.4.1 *Crystallization temperature*

To reduce the re-crystallization process described above a phase-change material with a higher crystallization temperature could be used. In this case the succeeding write pulses in the pulse train would cause less re-crystallization of the part of the amorphous mark formed by the previous write pulse. Thus, a broader amorphous mark would be formed. Choosing material with higher crystallization temperature would bring a few more advantages. Namely, improved shelf life and archival-life stability of the amorphous marks and reduced partial erasure of marks in the neighboring tracks on the disc (cross-erase effect). However, the use of materials with higher crystallization temperatures has several pitfalls. Most importantly, chalcogenide materials having higher crystallization temperatures usually also possess higher melting temperatures. Materials with higher melting temperature would require higher write powers, what inevitably leads to more heat accumulation in the stack.

5.1.4.2   *Crystallization speed*

Another trick to improve amorphous mark formation in an IPI stack is to choose materials with lower crystallization speed. The effect of material crystallization speed on the amount of re-crystallization can be verified by setting up the following experiment. The recording media are two discs with IPI stacks that have equivalent design in terms of absorption and cooling rate, but contain two phase-change materials that are different in terms of crystallization speed. A single write pulse of 4 ns is used to write amorphous marks at a constant linear disc velocity while varying the distance between the marks. In

Figure 88, normalized mark modulation (which is proportional to the mark area) is plotted versus the distance between the marks.

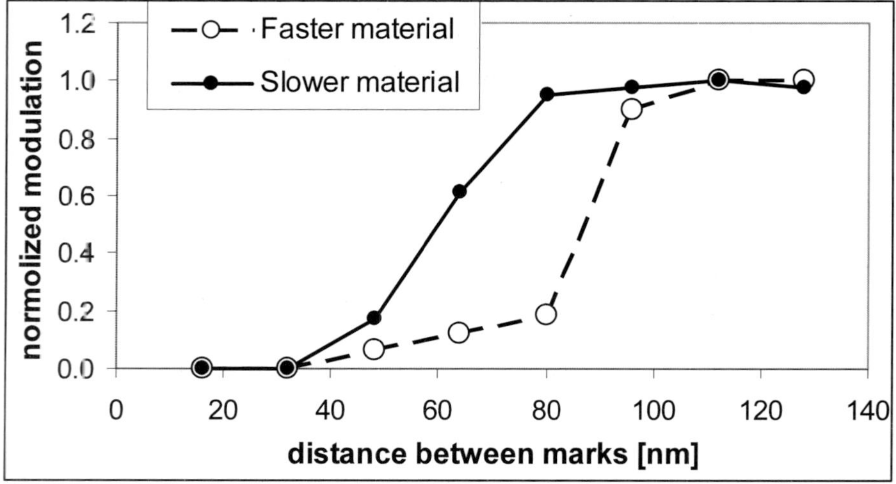

Figure 88. Normalized signal modulation versus distance between marks.

If the distance between the marks (*i.e.* between the write pulses) is very short each next-following write pulse causes complete re-crystallization of the amorphous mark written by the previous pulse. This results in zero signal modulation. With increasing distance between the marks (*i.e.* between the write pulses) the thermal influence of the next-following pulse on the previously formed amorphous mark decreases. As a consequence, less re-crystallization takes place and larger amorphous marks are left over leading to higher modulation. Ultimately, if the distance between the write pulses is big enough the temperature rise caused by a write pulse in the vicinity of the preceding mark is negligible. In this case no re-crystallization occurs and mark modulation reaches its maximum. Finally, for very long time between consecutive write pulses, separate marks will form and as a consequence the modulation will drop. For a given gap between the pulses the amount of re-crystallization is proportional to the crystallization speed of the material. As can be seen from the figure, despite the similar thermal behavior of the stacks higher modulation (*i.e.* larger marks) are obtained in the case of slower material. In other words, slower phase-change material exhibits less re-crystallization and, therefore, is more suitable for use in IPI stacks.

Despite the obvious improvement in terms of mark formation, materials with low crystallization speed cannot be applied in every optical storage system because of the imposed requirements on the data rate. It appears that currently available materials with crystallization rates slow enough to achieve good results in terms of marks size in IPI stacks are too slow to maintain a sufficient erasability rate during direct-overwrite process in the case of BD-system and higher speed versions of DVD-systems.

From the examples above it becomes clear that improving recording in IPI stacks by optimizing the phase-change material only is not efficient enough.

One more route to solve the mark re-crystallization problem during writing is to reduce the amount of heat that is accumulated in the stack during writing while maintaining sufficient heat for fast erasure. This can be done in two ways: by reducing the number of write pulses in the pulse train and by introducing a transparent heat sink in the recording stack.

5.1.4.3 *Write strategy*

Reducing the number of pulses in a write strategy results in that (i) less energy is pumped into the stack so that total heat accumulation is lower and (ii) distance between the subsequent pulses in the pulse-train is larger reducing the influence of the next-following write pulse on the part of the mark formed by the previous pulses of the pulse-train. The result of such a write-strategy optimization is shown in Figure 89.

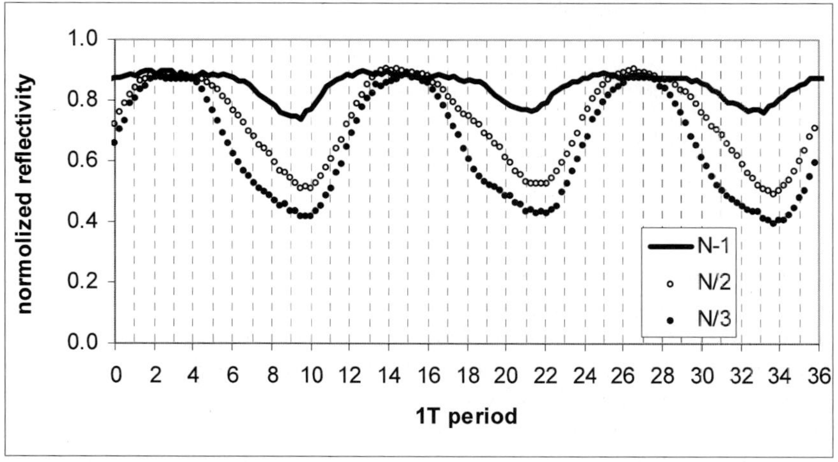

Figure 89. High-frequency signal of I6 carriers written in an IPI-type stack using N-1 (solid line), N/2 (open symbols) and N/3 (full symbols) write strategies.

In the plot, read-out signals are given for an I6-carrier where marks are written with five (N-1 or 1T strategy), three (N/2 or 2T strategy), and two (N/3 or 3T strategy) write pulses in the pulse train. As can be seen, the reduction of the number of pulses leads to an increase in signal modulation and to an improvement in the lengths of the marks and the spaces in the carrier. In Figure 90, the results of computer modeling and a TEM-image of an 8T-mark are displayed. The mark is written with three write pulses (N/3 or 3T write strategy). As can be seen, fewer write pulses in the pulse-train cause much less re-crystallization. However, as can be seen from the TEM picture, using a 3T write strategy for a slowly crystallizing phase-change material almost results in the formation of separated short marks.

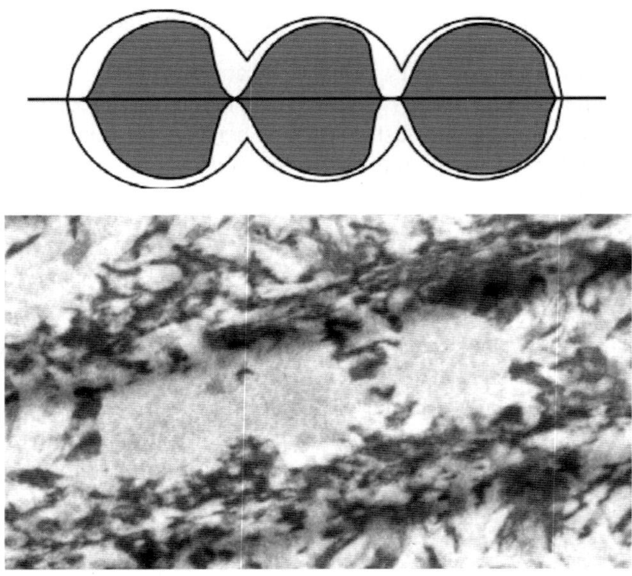

Figure 90. An 8T mark written with an N/3 (or 3T) write strategy in an IPI stack. upper panel - computer simulations, lower panel - TEM picture.

However, by reducing the number of pulses it becomes increasingly difficult to control the length of the marks/spaces. In Figure 91, the normalized lengths of marks and spaces are given as obtained with the N/3 write strategy. As can be seen, significant deviations occur in the case of shorter carriers. Applying different write-power levels and laser-pulse lengths for different mark lengths could minimize these deviations. Unfortunately, such a step would lead to a complicated write strategy and, as a consequence, will complicate the optimal power calibration (OPC) procedure in the drive and may compromise the disc playability.

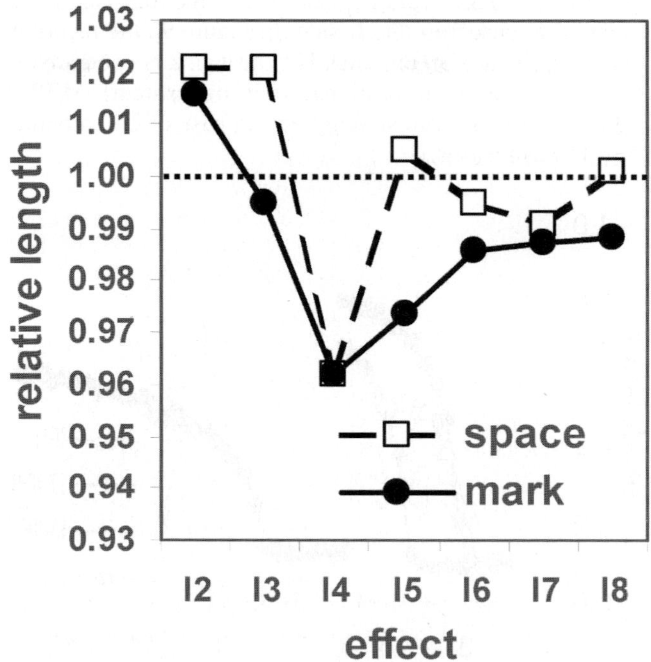

Figure 91. Normalized lengths of marks and spaces written in an IPI stack using a 3T (N/3) write strategy.

5.1.4.4 Stack design

Non-metallic heat sinks

An alternative solution to reduce the heat accumulation is to introduce a heat-sink layer into the recording stack. Keeping in mind the transmittance requirement, essentially transparent materials having relatively high heat capacity and conductivity would be beneficial. Generally, thermal conductivity/capacity of materials goes at the cost of optical transmittance. Therefore, a trade-off between thermal and optical properties of the heat-sink material has to be found. Analysis of a variety of materials including oxides, nitrides and oxi-nitrides of Al, Si, Ta, Ti, InSn, reveals that indium-tin-oxide (ITO) turns out to be one of the most attractive candidates. Among the materials considered ITO can provide the highest cooling rate and yet is sufficiently transparent to be applied in a semi-transparent optical stack. In the case of 405 nm laser wavelength, the most suitable stack design containing ITO as heat sink is $I_3TI_2PI_1$, where T is the transparent heat sink and the laser beam enters the stack from the I_1 side. In Figure 92 the recording performance of such a stack comprising a 30 nm thick ITO heat sink is compared to that of an IPI stack, IMIPI stack, and a conventional (not transparent) MIPI L0-stack. The recording is done with N-1 write strategy. In the case of IPI, modulation caused by the mark's "head" was measured.

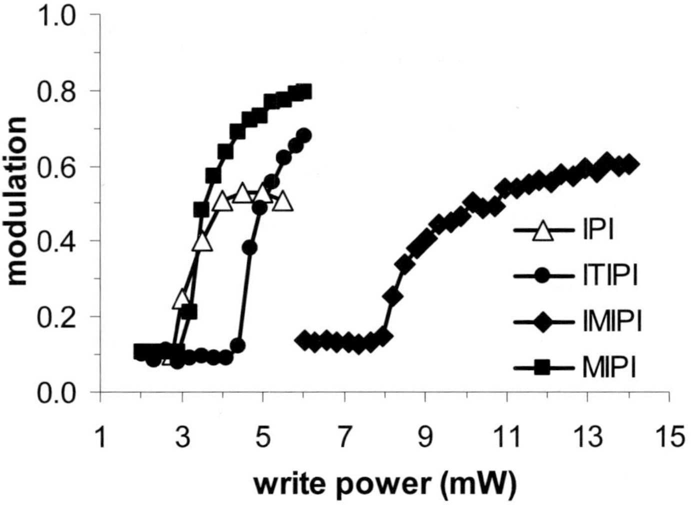

Figure 92. Recording performance of different stacks.

As can be seen from the figure, the presence of ITO increases the cooling rate in the stack and, as a consequence, improves the signal modulation in comparison to the IPI stack. However, from the broad analysis that has been done it appears that no non-metallic heat-sink material currently exists to adequately meet the requirements of phase-change optical recording at both DVD and BD conditions. Although the presence of such heat sinks in the stacks improves the cooling rate, this cooling rate is just not sufficient. Mark formation still suffers from re-crystallization during writing caused by heat accumulation. Recording with N-1 (or 1T) write strategy can only be realized if extremely short write pulses of high energy are applied. Such a regime is not desirable in terms of laser operation. Furthermore, mark erasure in stacks with non-metallic heat sinks is insufficient. Because of the poor thermal conductivities of the non-metallic materials, the stack is not pre-heated in front of the laser spot, and no complete re-crystallization (erasure) is achieved.

The effect of the thermal conductivity of the semi-transparent heat sink layer on the erase behavior is illustrated in Figure 93. In the figure, comparison is made of the instantaneous mark sizes during passage of the laser spot for three values of the thermal conductivity of the heat sink layer, namely λ_T=2, 5 and 10 W/(mK), to that obtained with the default IPI stack. The laser powers are adapted such that similar maximum temperatures were obtained for all recording stacks. It can be seen that a reasonable heat sink layer already causes a pronounced shift of the crystallization front.

Erasure ahead of the erasure spot ensures that the illuminated area is essentially crystalline. In the case of a semi-transparent heat sink, erasure ahead of the spot is less pronounced, which may result in an illuminated area that is still partially amorphous. In particular if the absorption of the crystalline and amorphous state is different, the associated temperature rise due to direct heating during erasure depends on the old data to be erased. This is probably the cause for incomplete erasure.

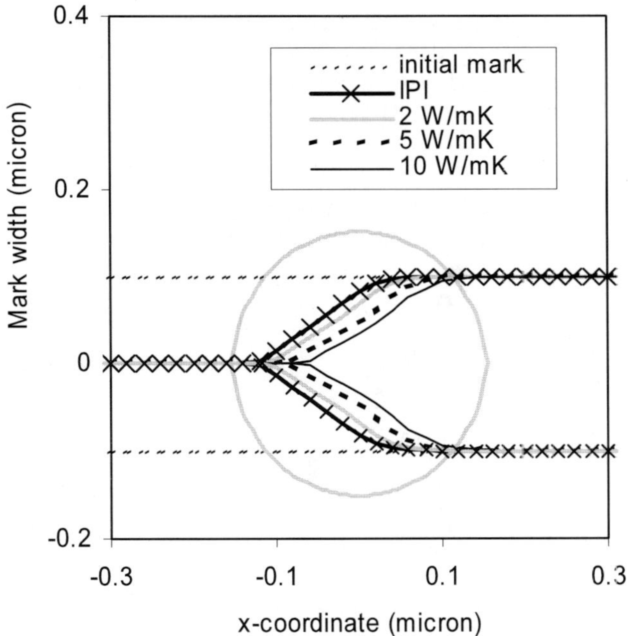

Figure 93. Calculated instantaneous mark shapes during DC-erase of a long carrier for different values of the thermal conductivity (λ_T=2, 5 and 10 W/(mK), BD conditions, LV=8.1 m/s). The dotted line indicates the initial mark edge. The gray circle denotes the 1/e-radius of the BD spot (R_0=151 nm).

Metallic heat sinks

From a variety of metals and their alloys that can be applied in an industrial optical disc manufacturing process, silver (or silver-based alloys) appears to be one of the most suitable candidates from both thermal and optical points of view. More specifically (i) this material has one of the largest differences between its refractive index and the refractive indices of other materials used in the recording stack giving rise to a high reflectivity level; (ii) at the wave length applied it has a relatively low absorption coefficient to allow for sufficient transmittance; (iii) it possesses high thermal capacity and conductivity resulting in high cooling rates in the recording stack. The disadvantage of silver as a material for a heat-sink layer in a phase-change recording stack is that it is very susceptible to sulfur, which is often used in the dielectric layers of the stack and its tendency to exhibit island- like morphology when used as a thin layer. Alloying silver with certain elements can improve the corrosion resistance but it affects thermal performance and morphology of the layer. In Figure 32 optical parameters of a silver layer at the 405 nm wavelength are given as a function of the layer thickness.

As can be seen from the plot, the thinner the layer the higher the refractive index of the material. This means that optical characteristics of a recording stack comprising a thin Ag layer will strongly depend on the Ag layer thickness. Optical analysis and recording experiments reveal that the most efficient design for a semitransparent BD recording stack is that of the $I_3MI_2PI_1$-type, where the laser beam enters the stack from the I_1 side.

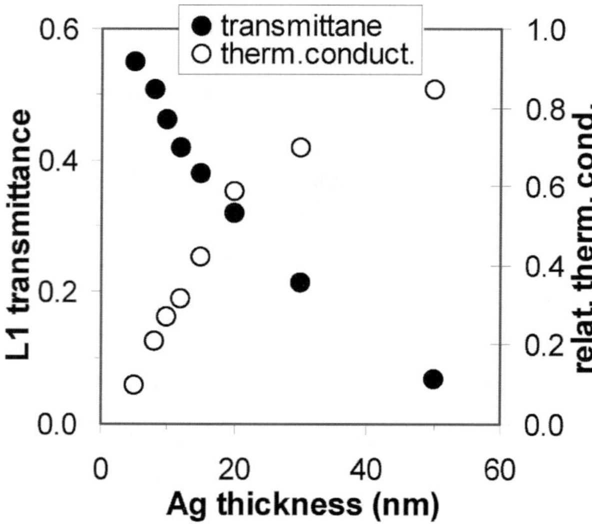

Figure 94. Characteristics of an IMIPI stack as a function of the Ag heat sink layer thickness.

In Figure 94 both the transmittance of such a stack and the relative thermal conductivity of the heat-sink layer are plotted versus the heat-sink thickness. The unity of relative thermal conductivity corresponds to a 120 nm thick silver layer as used in a conventional recording stack. As can be seen from the figure, a transmittance of about 50% can be obtained at the silver layer thickness of about 10 nm. At this point the relative thermal conductivity is approximately 20%. Thermal calculations reveal that although re-crystallization of amorphous marks in such a stack caused by the consecutive write pulses is higher compared to a stack with a 120 nm thick silver layer, this re-crystallization is much smaller compared to the case when a non-metallic heat sink is used. In Figure 95 mark shapes obtained by computer simulations for an IMIPI semitransparent stack and a conventional MIPI stack are shown. The thickness of the Ag heat-sink layer in the stacks is 10nm and 120 nm, respectively. In the simulations, the N-1 (or 1T) write strategy was used. As can be seen, only a small difference in re-crystallization can be observed.

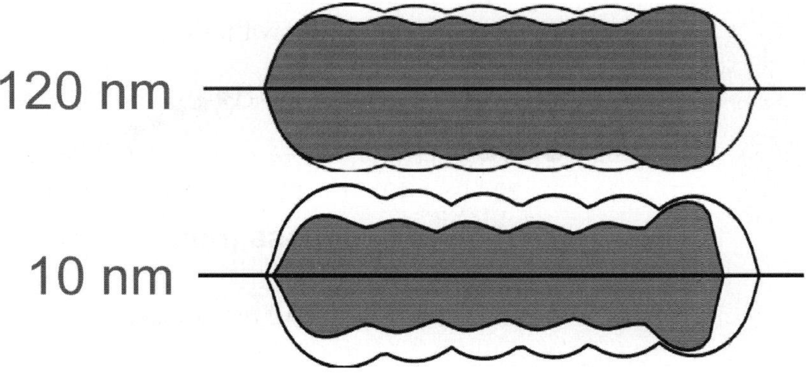

Figure 95. Mark shapes obtained by computer simulations for IMIPI and MIPI stacks.

It is worth to mention here the role of the third dielectric layer I_3 placed behind the metallic heat-sink/reflector layer in the stack. In Figure 96 the calculated transmittance and reflectivity-times-contrast parameter of semi-transparent stacks with and without I_3-layer are given. When the I_3-layer is absent the reduction of the silver layer thickness down to 10 nm is not enough to obtain 50% transmittance. However, further reduction of the silver layer thickness will deteriorate both thermal performance of the stack and its reflectivity and contrast levels. By placing a dielectric layer behind the silver layer a substantial transmittance gain can be achieved without affecting the reflectivity-times-contrast parameter.

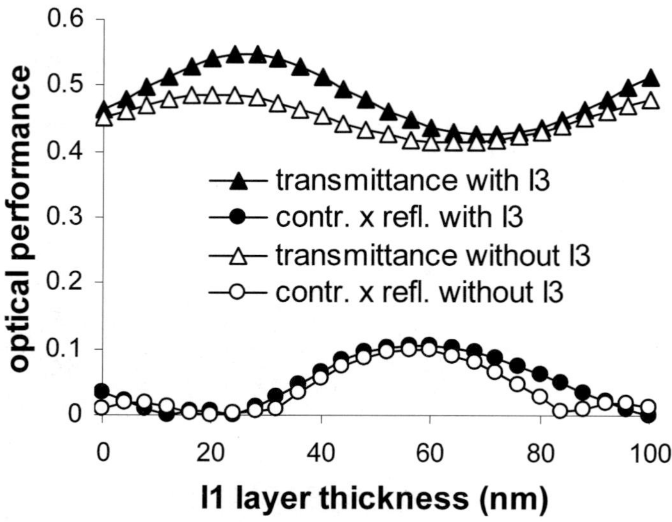

Figure 96. The role of I3 layer in the $I_3MI_2PI_1$ stack design.

5.1.5. *Transmittance difference*

An interesting and important issue related to dual-layer media and recording is the dependence of transmittance level of the semi-transparent L1-stack on its recorded state. This issue is caused by the fact that the crystalline and the amorphous state of phase-change materials have different transmittance (see Figure 97). For the semitransparent recording L1-layer an effective transmittance close to 50% is required. Assuming that the amorphous marks' width is half the track pitch and considering groove-only recording, the effective transmittance of the recorded L1-layer can be calculated as $T_{eff} = \frac{3}{4}T_c + \frac{1}{4}T_a$, where T_c and T_a are the crystalline and amorphous transmittances of the stack, respectively. (In this simplified picture the effect of diffraction is omitted). If L1 is empty its transmittance equals the crystalline transmittance. How well transmittance difference between L0 and L1 can be balanced depends on the type of phase-change material used. Figure 99 shows that the transmittance difference between the amorphous and crystalline state is larger in the case of growth dominated phase-change materials. For this reason, it is difficult to achieve transmittance balance. However, transmittance-balanced semitransparent stacks can be designed in the case of NDM type of phase-change materials. [124] In such stacks, the transmittance of the empty and the written area of the layer is balanced by optimizing the composition of the phase-change material and by tuning the optical design of the recording stack.

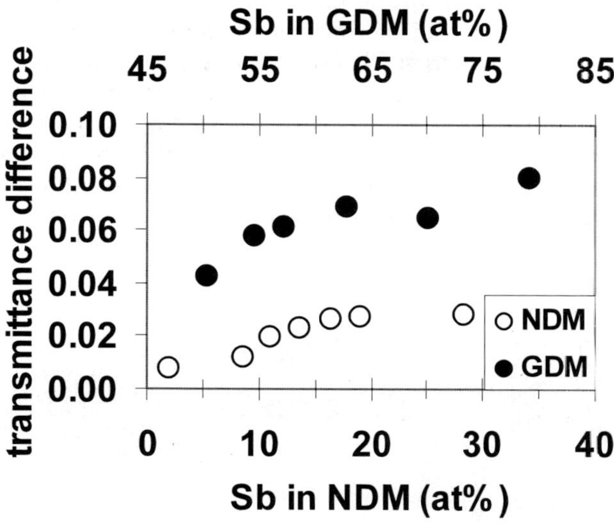

Figure 97. Transmittance difference between amorphous and crystalline states for NDM and GDM classes of phase-change materials plotted versus Sb concentration.

When the laser beam is focused onto the deeper laying L0-layer, the size of the laser spot on the L1-layer in the case of BD system is about 30 μm in diameter, so it covers about a hundred recording tracks. If the L1-layer is partly written, areas with different effective transmittance are present and a transmittance change occurs when the laser beam scans across the disc. This leads to variations of the intensity of the laser beam reaching the L0-layer as well as variations in the intensity of the reflected light when the L0-layer is read-out. As a consequence, disturbances in the servo-signals, increase in jitter and in symbol error rate (SER) may take place.

Depending on the way the data marks are placed in a partly recorded L1-layer two types of transition across the written-empty borders are of interest. One is when the transmittance change occurs slowly with a period of one disc revolution. This can happen if a band is written on the L1-layer full-way the disc circumference (see panel (a) of Figure 98). When the laser beam is focused through L1 onto L0, the laser spot on L1 can fall partly on the written band and partly on the empty tracks of L1. Due to the presence of mutual eccentricity between the groove structures of L1 and L0 the amount of written/empty areas of L1 covered by the spot will vary slowly with a period of one disc revolution. As a consequence, the effective transmittance of L1 will vary and lead to variations in the intensity of the laser beam reaching and reflected from L0.

The second type of transition is when the transmittance change occurs very quickly and to its maximum extent. This occurs when the transmittance of L1 changes abruptly in the tangential direction of the tracks being followed and the width of the written area is larger than the size of the laser spot (see panel (b) of Figure 98). It should be stated that this kind of situation will hardly occur in reality. In real applications the data is written in blocks of a certain length. In the case of BD, for instance, two to five blocks can fit one disc circumference depending on the track radius on the disc. If the tracks written on L1 form a band with a width of about the size of the laser spot when the laser is focused on L0, the leading and trailing edges of the blocks can hardly form a straight line along the radial of the disc. Thus, the second type of transition represents an artificial worst-case situation.

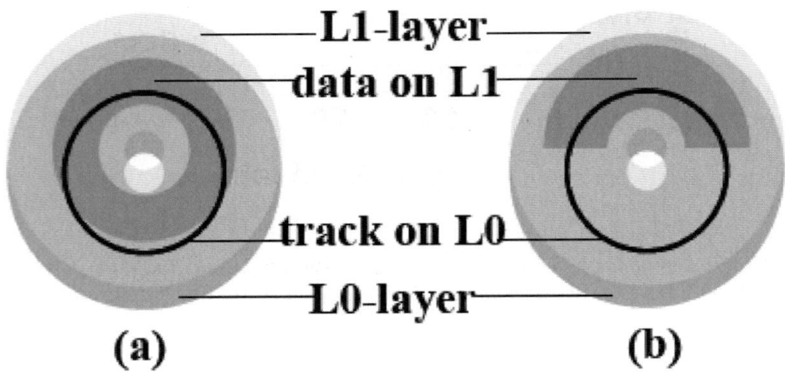

Figure 98. Two possible transitions across the written/empty border. In the figure, the mutual eccentricity between L0 and L1 is exaggerated.

To look at the influence of the two transition types in the L1 layer on the performance of the L0-layer a special BD dual-layer rewritable disc was designed. The L1 stack was made to have its transmittance changing slightly along the disc circumference. This would mimic the first transition type as far as addressing of the L0 layer is concerned. Then, 2500 tracks were written on the L1-layer with random data halfway of the circumference. As a result, a half-doughnut-shaped written band with a width of 800 μm was created. The relative transmittance difference between empty and recorded state of the L1-layer was about 4.5%. The L0-layer was subsequently recorded and read-out along the middle line of the band. Figure 99 shows the resulting high-frequency signal. A jump in L0 reflectivity caused by the transmittance difference between the written and empty areas of L1 can be observed and amounts to about 9.5% relative value.

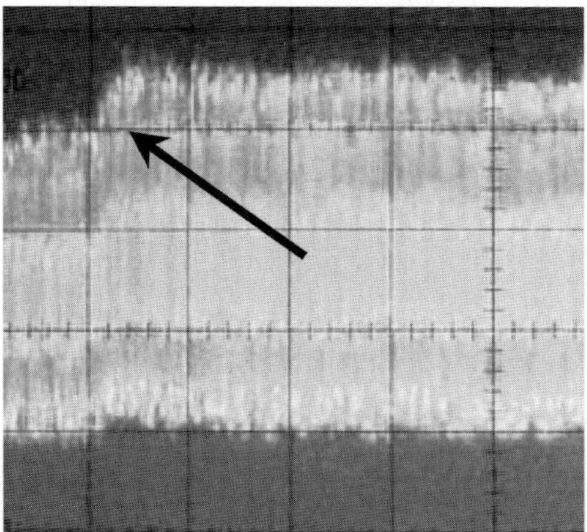

Figure 99. High-frequency signal of the L0-layer. The L0-layer was written with random data and read-out through the half-doughnut-shaped data pattern of the L1-layer. The jump in signal level is caused by abrupt change in the effective transmittance of the L1-layer.

In Figure 100, high frequency, residual focus error, and residual tracking error signals of an empty track on L0-layer are given for one full disc circumference. The L0-layer signals are acquired through the half-doughnut-shaped data pattern of the L1-layer. An abrupt jump and a gradual variation in the reflectivity can be observed, which do the two types of transmittance transition cause. As can be seen from Figure 100, the 9.5% relative reflectivity jump does not disturb the servo performance of the system.

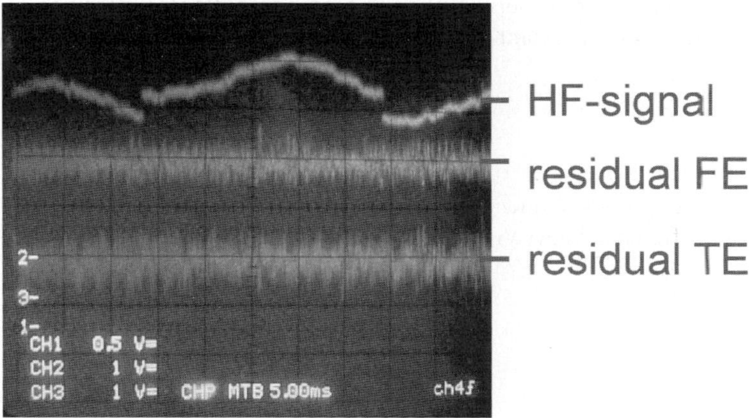

Figure 100. High frequency, residual focus error, and residual tracking error signals of an empty L0-track read-out through the half-doughnut-shaped data pattern of the L1-layer.

In Figure 101, L0 signal modulation versus write power is plotted. The open and full circles in the plot correspond to L0-layer recorded through empty L1 or written L1, respectively. As can be seen from the figure, the transmittance difference of the L1-layer has hardy any influence on the write power of the L0-layer for sufficiently high modulation values.

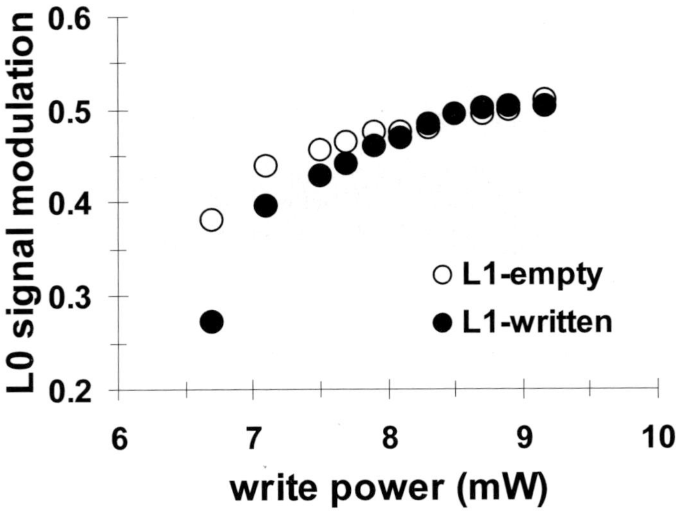

Figure 101. Signal modulation of the L0-layer recorded through empty and written L1-layer.

The 4.5% relative transmittance change in the L1-layer has also no influence on the jitter and SER of the L0-layer.

Although in the example considered in this section no significant disturbances in system performance were observed, the issue of transmittance difference in dual-layer media remains important since it is specific for a particular media-drive combination.

5.2.   *Media for high-speed phase-change recording*

5.2.1.   *Introduction to speed race*

The data recording speed is an important performance factor in optical phase-change recording. Figure 102 shows the recording speeds that have been standardized and/or demonstrated for the formats of CD, DVD and BD by the year 2003. In 2002, ultra-speed 32× CD-RW has been standardized (1× CD corresponds to 1.2 Mbit/s or a linear velocity of 1.2 m/s). The high-speed DVD+RW standard has been updated in 2004 to comprise recording at 3.3-8x DVD (1× DVD corresponds to 11 Mbit/s or a linear velocity of 3.5 m/s) and for BD 2× has been standardized (1× BD corresponds to a data rate of 36 Mbit/s or a linear velocity of 5 m/s). Although the data transfer rates for the three CD, DVD and BD generations vary substantially due to the difference in data density, the maximum achievable linear velocity, measured at the outer radius of the disc, compares quite well. At present, the achievable maximum linear velocity of an optical disc is 56 m/s. Above this velocity, a number of issues arise, such as the mechanical stability of the polycarbonate substrate, robustness of the servo-system, acoustic noise level, power consumption etc.

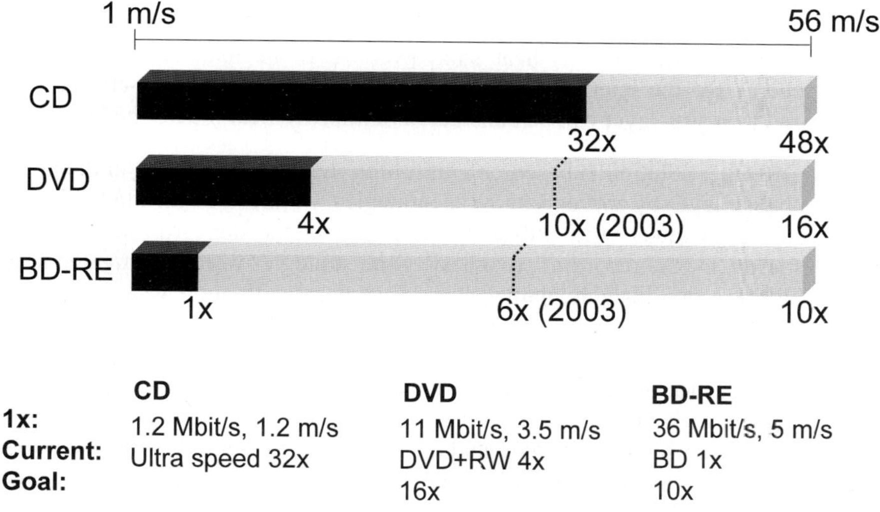

Figure 102. Demonstrated speeds for the CD, DVD and BD optical disc formats as of 2003.

The main challenge of high-speed phase-change recording is of fundamental nature. Direct overwrite requires that old data are erased in the same passage of the laser spot in which new data are written. The consequence of this is that at high recording speeds the time for erasure of old marks is inversely proportional to the recording speed. To enable complete erasure of old marks at high recording speeds, the crystallization speed of the materials needs to be increased. Fast crystallizing materials, however, may cause problems during data recording. As has been explained in Chapter 4, a mark of certain length consists of overlapping amorphous areas, which are created by consecutive write pulses. A laser pulse fired to form the next part of the mark causes heating up of the mark made by the previous write pulse. Both this direct heating and heat diffusion through the stack causes unwanted re-crystallization of the mark being written. The challenge of ultra-speed phase-change recording is to find a delicate compromise between fast crystallization of old data during direct overwrite and sufficient suppression of crystallization during write of new data.

Two solutions can be anticipated:

1. An obvious solution is to pursue a fast-cooling recording stack in which the dissipated heat quickly diffuses away after application of the write pulse. The recording stack is then less sensitive to the accumulated heat. This can be achieved by application of fast-cooling materials such as silver or other metallic conductors as heat sink layer. Also the material and thickness choice for the interference layer situated between the phase-change layer and heat sink layer can be considered to improve heat loss to the heat sink.

2. The other solution is based on controlling the heat dissipation in the phase-change layer by appropriate write strategies. Control of heat can be done by shorter or fewer write pulses. In the course of the standardization of ultra-speed CD-RW recording, two basic write strategies were investigated based on these two insights, namely the 1T strategy with short write pulses and the 2T strategy with a reduced number of write pulses. In a 2T write strategy (WS), one write pulse is used for every two-clock cycles. A consequence of this WS scheme is that consecutive even and odd marks are written with the same number of write pulses. For example, three write pulses are used to write a 6T and a 7T mark. The essential point of the write strategy is the choice of the parameters that define the difference between even and odd marks. Even marks are defined in a straightforward way by a pulse-train with multi-pulse length (T_{mp}) and cooling gaps in between the pulses (T_c). The concept of the 2T write strategy originates from Ricoh [117] and MCC. [118]

5.2.2.   *Materials for high-speed recording*

The rate-limiting step in rewritable optical recording is the speed at which previous amorphous marks can be re-crystallized (erased) during direct overwrite. Therefore, for high data rate recording, the phase-change material should be optimized. Increasing the crystallization rate of the phase-change material, however, may not be done at the expense of other materials properties, such as optical constants, amorphous phase stability and media noise. Especially the combination of high (laser-induced) crystallization rate and high amorphous phase stability is often conflicting, although extremely important. Namely, spontaneous re-crystallization of amorphous data marks at room temperature or slightly elevated temperature will lead to an increase in bit error rate.

Three main classes are currently used for phase-change recording, $Ge_2Sb_2Te_5$ compositions, doped-Sb_2Te compositions and doped-Sb compositions. The crystallization process of the latter two is mainly dominated by fast crystal growth, while $Ge_2Sb_2Te_5$ compositions posses a strong nucleation-dominated behavior. The difference in maximum possible data rate between nucleation and growth-dominated materials is illustrated in Figure 103. [119] Shown are the maximum user data rates as a function of the reciprocal spot size for a nucleation-dominated $Ge_2Sb_2Te_5$ compositions and a growth-dominated doped-Sb_2Te composition. The spot size is taken as the $1/e$ radius (R_0) of the focused laser beam and is given for the four optical storage systems, namely CD (R_0) =500 nm, DVD (R_0=350 nm), HD-DVD (R_0=245 nm), and BD (R_0=151 nm). It can be seen that the maximum user data rate remains more or less constant for the nucleation dominated material class, while it strongly increases for the growth-dominated materials. This can be understood from the two different mechanisms of crystallization. In case of nucleation, stable embryonic clusters evolve into stable nuclei, which grow steadily in the radial direction. The amorphous mark is completely erased if all amorphous area is re-crystallized from the numerous stable nuclei inside the mark. This process seems to be independent on mark size. Or, it may even become worse for smaller spot if the somewhat steeper temperature gradients in case of a higher density (thus smaller spot) are taken into account. On the contrary, re-crystallization for the growth-dominated materials starts from the amorphous-crystalline interface of the mark and is pointed inwards. The time required for complete re-crystallization is, therefore, determined by the distance to be overcome and, thus, by the radius of the mark.

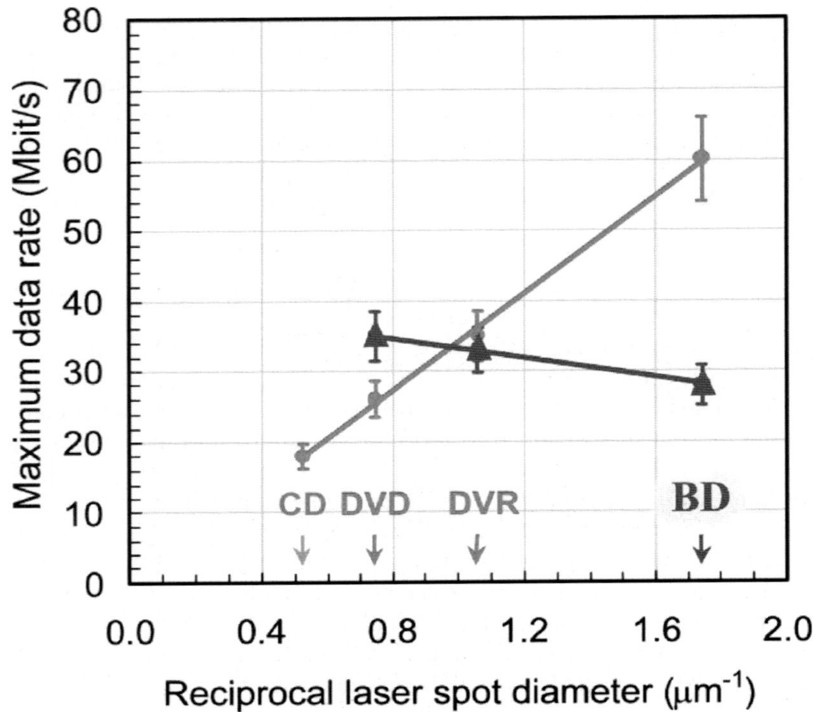

Figure 103. Maximum user data rate as a function of the reciprocal spot size of four consecutive optical recording systems for a nucleation-dominated $Ge_2Sb_2Te_5$ composition (solid circles) and a growth-dominated doped- Sb_2Te composition (solid triangles). DVR 'red' refers to the high-NA system (NA=0.85) based on a red laser, BD refers to the current BD system. [119]

For higher data density systems, the spot size becomes accordingly smaller, hence, leading to a reduced erasure time and a higher maximum data rate. For this reason, benefits can be expected when using growth-dominant phase-change materials to achieve high data rates in high-density systems. In the experiments described in this section, growth-dominant phase-change materials are hence used as the recording material.

5.2.3.   Optical and thermal design of high-speed recording media

The optical design of high-speed recording media is not different from the design of lower-speed discs. A typical stack design for a DVD+RW is given in Figure 104. The phase-change layer is sandwiched between two dielectric layers and is separated from a metal layer that acts as a mirror and heat sink. The thickness of the layers can be adjusted to arrive at a proper reflectivity, preferably above 20% for the DVD-format, and a proper contrast between the amorphous and crystalline phase, ideally over 95%. The stack layout for high-speed CD and BD discs is quite similar, apart from slight differences in the applied materials and the layer thickness.

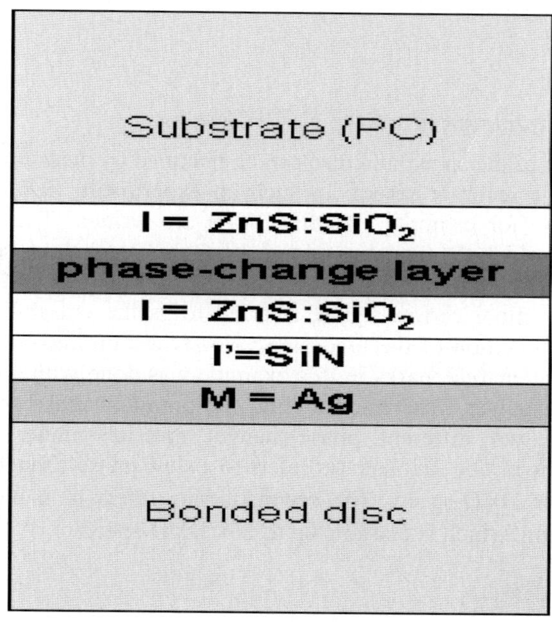

Figure 104. Example of a phase-change recording stack used for high-speed DVD+RW recording.

However, to be able to write amorphous marks in a fast crystallizing phase-change material, the stack should be thermally designed such that the heat of the write pulses can diffuse away quickly. This implies that for the metal mirror only materials can be used with a high thermal conductivity, such as Ag. Also, the thickness of this Ag-layer is increased with respect to low-speed discs, and is typically between 150 and 200 nm. Furthermore, the interface layer between the phase-change and the metal layer is optimized to improve the heat conduction to the heat sink. This is typically done either by decreasing the thickness of the dielectric interface layer or by choosing a dielectric material with a higher thermal conductivity.

5.2.4.   Characterization of high-speed discs

In this sub-section, methods and methodologies are described that are used to characterize high-speed phase-change discs. The maximum recording speed at which data can be directly written in a rewritable disc is determined by the crystallization speed of the phase-change material. Re-crystallization during write, the so-called back-growth, limits the achieved modulation and needs to be controlled via appropriate write strategies.

5.2.4.1. Crystallization speed

A good indication of the speed of a disc can be obtained by measuring the maximum erase velocity at a recorder set-up. In such an experiment, first, carriers of long amorphous marks (for example 14T for DVD) are written. Then, the maximum linear velocity at which the signal amplitude of these carriers can be reduced by 25 dB is determined. An example is given in Figure 105. The ratio of the carrier signal after erasure and initial carrier signal, referred to as the erasability, is plotted in terms of dB as a function of the recording velocity (recording = erasing velocity). Erasure of the written 14T marks in this example was done with a continuous laser power. The erase power was varied to find the optimum erase conditions. In the figure, data for two different phase-change materials under optimum erase conditions are shown. The first data set refers to a slow phase-change material that is erasable up to 4× DVD-speed. The second series refers to a high-speed phase-change composition, which is erasable up to 16× DVD-speed.

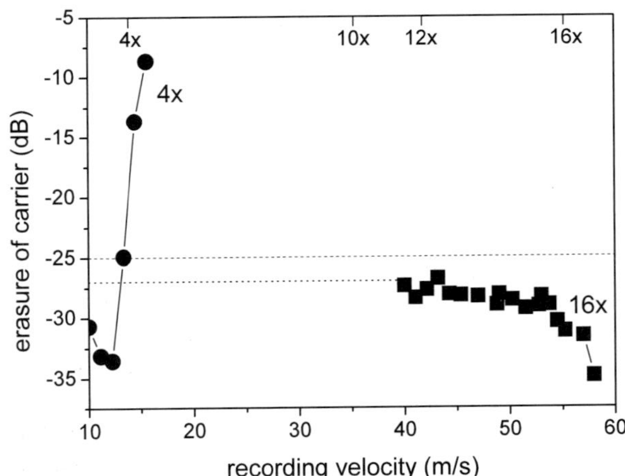

Figure 105. Erasability of long 14T carriers written in a DVD+RW disc as a function of the recording velocity. Shown are two sets of data, one set refers to a slow composition (4×), the other series refers to a fast phase-change composition (16×).

The shape of the experimental curves can be understood from computer simulations of the erasure process. Figure 106 shows the results of calculations of the mark width as a function of the continuous erase power for a slow and a fast crystallizing phase-change material. The erasure was done with a single passage of the laser spot. While complete erasure is achieved at a medium erase power (2 mW) in case of the fast phase-change composition, a residual mark remains for the slow phase-change composition. An increase of the erase power leads to melting of the phase-change layer and subsequently to a further amorphisation of the phase-change layer (an amorphous trace remains in the recording layer). Note that the calculations in Figure 106 are performed for an IPI-stack, *i.e.* a recording stack without a metal heat sink. Simulations with a stack with heat sink give similar results, but the required erase powers to achieve complete erasure are significantly higher because of the reduced power sensitivity.

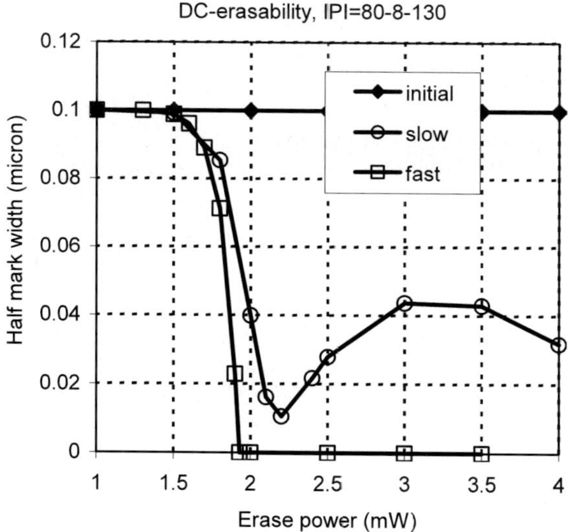

Figure 106. Computer simulations of (partially) re-crystallized amorphous marks in a fast and a slow phase-change composition.

5.2.4.2. Back growth

If the heat generated in the phase-change layer cannot dissipate into the heat sink sufficiently fast, re-crystallization of already amorphised areas may occur. In that case, the laser pulses intended to locally melt the phase-change layer will also heat the surrounding area and induce re-crystallization of the marks just written. For fast-crystallizing phase-change materials, this effect can be particularly severe. The effect of material crystallization speed on the amount of re-crystallization can be visualized very well by performing computer simulations. The results of such simulations are given in Figure 107. The solid line represents the edge of the area molten during recording. The shaded area represents the final mark size. As can be seen, if the stacks are made thermally equal and the same disc velocity is used, the molten area is the same for the slower and the faster phase-change material. However, the final mark size left after partial re-crystallization is larger in the case of slower material.

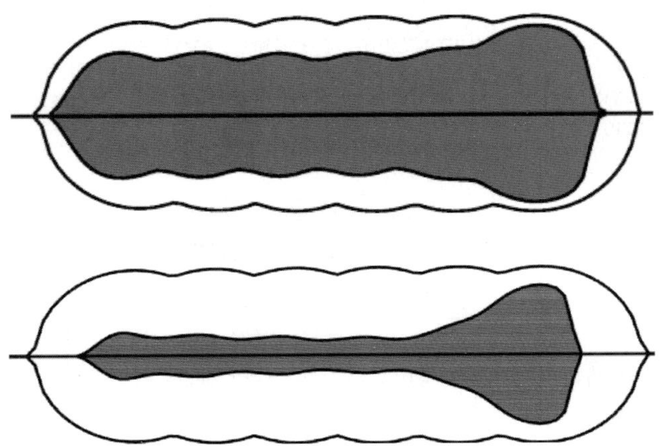

Figure 107. Mark shapes obtained by computer simulations of a recording process. The solid line represents the melt edge; the shaded area shows the mark size left after recording. (a) - material with low crystallization speed (v_{max}=4.8 m/s) (b) - material with high crystallization speed (v_{max}=8.1 m/s).

It is, therefore, important to measure the amount of back growth in a phase-change stack. The so-called modulation-reduction measurement was specially developed to characterize a phase-change stack. In such an experiment, long (*e.g.* 11T) carriers are written by an increasing number of laser pulses and the resulting modulation is measured. This is schematically shown in Figure 108. When the time between the pulses becomes too short, amorphous area will be partly re-crystallized by the subsequent laser pulse and a reduced modulation is measured. Figure 109 shows the modulation-reduction for discs designed for 6×, 10× and 16× DVD-speeds. The maximum modulation is normalized to 1. It is clear that high-speed discs need more time between two write pulses to prevent reduced modulation due to re-crystallization. In other words, high-speed discs show more back growth. We will use the time between pulses at 90% of the maximum modulation as a measure for the amount of back growth.

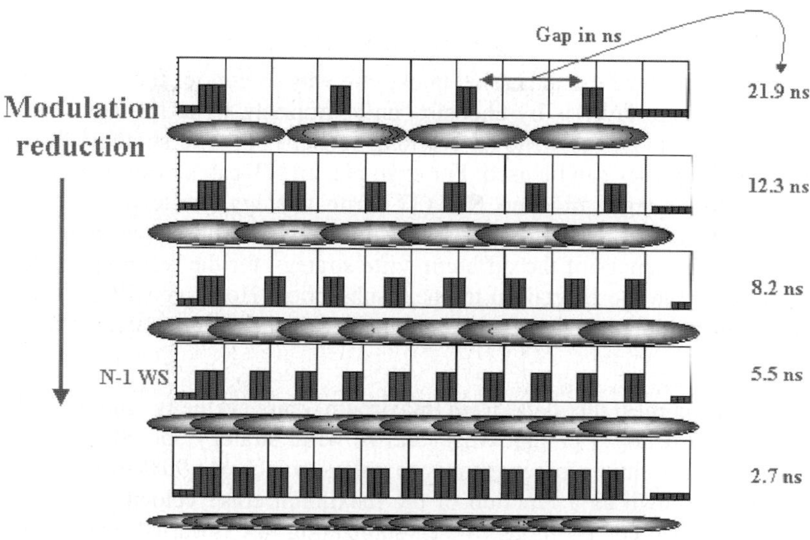

Figure 108. Schematic of a modulation-reduction experiment to characterize the writing behavior of marks in a phase-change stack. The panels with rectangular blocks represent schematically the laser power modulation profiles used to write the marks. The ovals in the figure represent amorphous dots created by the write pulses.

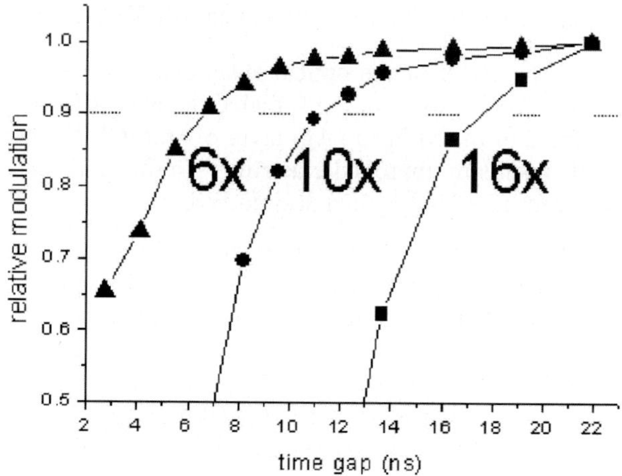

Figure 109. Modulation reduction results for a high-speed DVD disc at 6×, 10× and 16× recording velocity.

5.2.4.3. Write strategy

As has been mentioned earlier in this book, one way to control the heat dissipation in the phase-change layer is by choosing an appropriate write strategy. Instead of using (N-1) write pulses to write an NT-long carrier (1T write strategy), a reduced number of write pulses can be used. For example, an NT-carrier can be written with N/2 (2T write strategy) or even N/3 (3T write strategy) write pulses. Also, by reducing the length of the write pulses the time between subsequent pulses can be increased. Consequences of the different write strategy for the resulting mark shape will be discussed in more detail in the next subsection. Here, we will only focus on the maximum modulation that can be attained with the write strategy for a given disc.

Figure 110 combines the data from maximum erase velocity and modulation-reduction experiments to predict what kind of write strategy should be used for a particular disc. The time gap between write pulses (to achieve 90% of the maximum modulation) is shown as a function of the maximum erase velocity for Ge-Sn-Sb based phase-change compositions (the crystallization rate is varied by changing the Ge concentration). The gap is divided by the period T (T=9.55 ns at 4×), resulting in a parameter that is independent of the write velocity. In agreement with Figure 109, it is observed that back growth increases for discs with higher maximum erasability.

In Figure 110, dotted lines representing write strategies are drawn. The lines indicate 1T-period-based, 2T-, and 3T-period-based strategies. The length of the write pulses in this example was 3 ns for all write strategies. In order to record a disc with a given write strategy (WS), the coordinates of gap and erase velocity should be below the corresponding WS-line. For the phase-change stacks in Figure 110(a), recording up to about 6× DVD is possible with a 1T WS (with 3 ns pulses), recording up to 10× DVD is possible with a 2T WS, and at higher speeds 3T WS should be applied.

Figure 110(b) shows the effect of the optical stack design on erasability and back growth. By decreasing the thickness of the I_2 dielectric layer $((ZnS)_{80}:(SiO_2)_{20})$ or by increasing the thickness of the M-layer (Ag) layer of an MCI_2PI_1 stack back growth can be reduced. This results in improved erasability. In this manner, it is possible to design a disc suitable for recording with a desired WS.

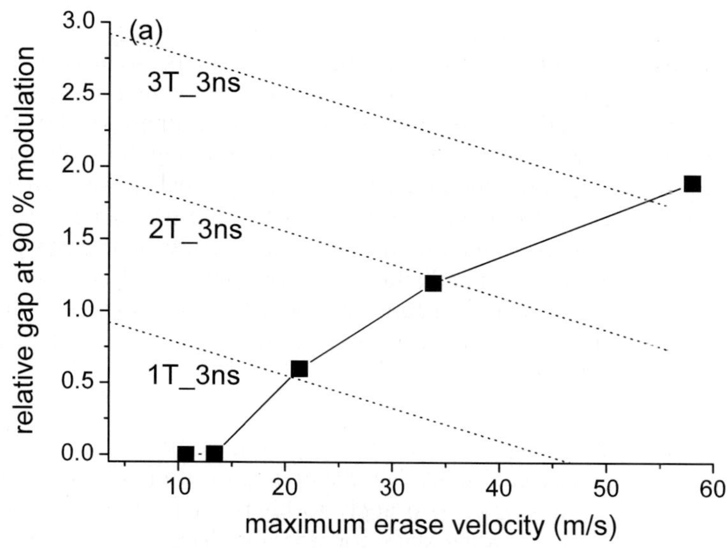

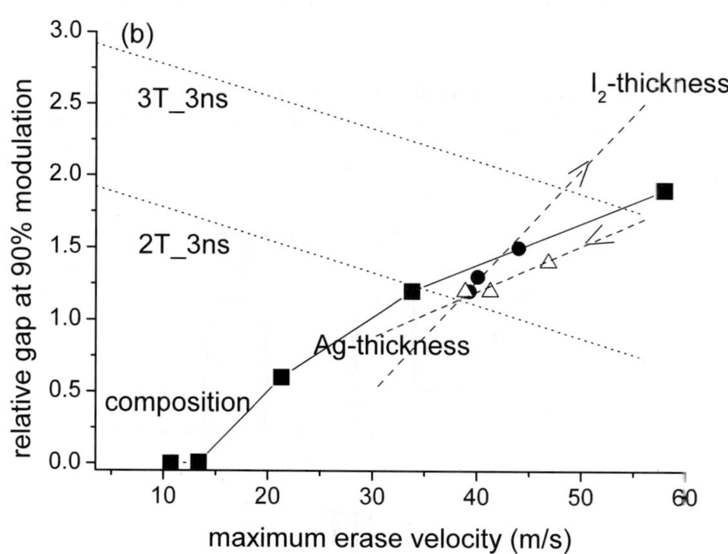

Figure 110. Plot of the relative cooling gap as a function of the erase velocity.

5.2.5.   *Mark formation in ultra-high speed recording stacks*

Heat accumulation in high-speed recording stacks can be so severe, that mark formation is hindered by unwanted re-crystallization. In section 5.2.1, it was mentioned that mark formation in high-speed discs can be improved by 1) a faster cooling structure so that the temperature drops quicker during the cooling periods in the laser pulse train 2) an appropriate write strategy. Such a write strategy may consist of short laser pulses (for example the 1T-16× CD-RW write strategy in the Ultra-speed Orange Book, version 1.0) or a reduced number of write pulses (for example the 2T write strategies in the same Orange Book). In this section the effects of the write strategy on mark formation will be discussed. Computer simulations, used to understand the mark formation process in more detail, are compared to TEM measurements of experimental data.

Mark formation @ 2T strategy
CD-RW results are considered to explain the principles of mark formation in high-speed discs but these principles also apply to high-speed DVD+RW and BD-RE recording. The CD-RW recording stack comprises a phase-change layer (P, doped SbTe), two zinc-sulphide-quartz layers (I) and a metal heat sink layer (M). The I_1PI_2M recording stack had a I_1PI_2M=90-16-25-100 nm layer thickness distribution. A 0.5 numerical aperture and a 780 nm laser wavelength characterize the CD-RW system. The theoretical 1/e radius of the focused spot reads R_0=494 nm. The computer simulation results were generated with the mark-formation tool as discussed in chapter 3.

First, the 2T write strategy is considered. A mark of NT length is written with N/2 write pulses. The pulse train used to write the leading part of the marks is given in Figure 111.

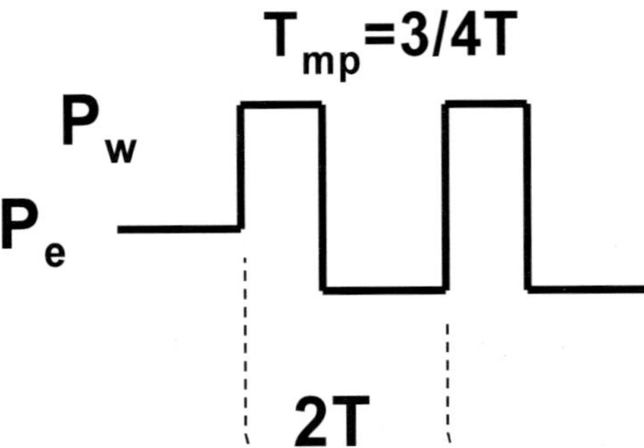

Figure 111. Head of the 2T write strategy to write long marks at high speed CD-RW (in this particular case 32x). P_e denotes the erase power, P_w denotes write power, and Tmp is the length of the write pulse.

The length of the write (multi) pulse is indicated with T_{mp}. The subsequent write pulse is fired 2T later in time. The cooling gap between the multi-pulses is then $T_{cool}=2T-T_{mp}$. For longer marks, this write-pulse cooling-gap sequence is repeated. Each laser pulse is used to write a 2T partition of the total mark. Figure 112 gives pulse strategies for the trailing part of the even (left panel) and odd (right panel) amorphous marks. For writing even marks, the last cooling gap is reduced with a factor θ. For odd marks, the last write pulse is extended with Δ to melt an elongated part of the track. To suppress re-crystallization, the extended pulse is also shifted by $\Delta 1$. The cooling gap reduction is then θ_{odd}. The relevant parameters are tabulated in Table 9 for various CD-RW recording velocities. The pulse time T_{mp} was experimentally determined and was constant for all recording speeds (T_{mp}=5.4 ns = 0.75T at 32×). Since the physical mark length (recorded in the disc) should be equal for all recording speeds, the clock time is accordingly reduced with increasing linear velocity. The elongation $\Delta 1$ was constant in ns ($\Delta 1$=3.6ns).

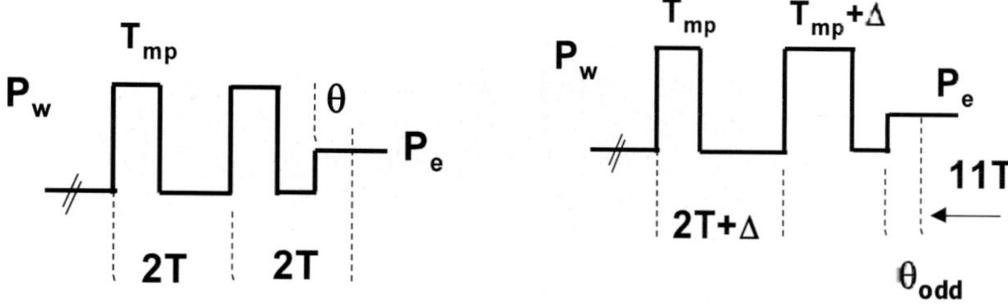

Figure 112. Schematics of the tail of the 2T write strategy to write long marks at high speed CD-RW for an even (left panel) and odd (right panel) mark length. Δ is the pulse length and cooling gap elongation to fill the 3T remainder of the odd mark length.

Table 9. 2T write strategy parameters used in the thermal simulations.

×-factor	Linear velocity (m/s)	Clock time (ns)	Pulse time (ns)	Elongation θ (T)	θ_{odd} (T)
8×	9.6	28.87	5.4 (3/16 T)	¾	1
16×	19.2	14.44	5.4 (3/8 T)	¾	1
32×	38.4	7.22	5.4 (3/4 T)	¾	1
40×	51.2	5.78	5.4 (≈ T)	¾	1

Computer simulation results of a 10T mark written with the 2T strategy are given in Figure 113 for four different recording speeds. The corresponding temperature-time responses monitored in the center of the mark are shown in Figure 114. The solid black line in Figure 113 indicates the melt-edge; the gray area denotes the remaining amorphous area after partial re-crystallization. Two distinct phenomena are clearly observed: 1) the re-crystallization in the tail of the mark due to the laser-induced heating by the applied erase power following the pulse train 2) the serrated side edge, which is caused by partial re-crystallization during writing of the amorphous areas.

The mark written at the lowest recording speed (8x) shows distinct re-crystallization at the leading edge. The rather high erase power of 8mW, taken constant for all recording velocities, caused a high temperature bias level at the low recording velocity (see the time-responses in Figure 114). The temperature response of the first write pulse is superimposed on that DC-response and leads to a broad melt-edge (see mark-shapes in Figure 113). The consequent accumulated heat also causes re-crystallization in the cooling-down phase, in between the first and second laser pulse.

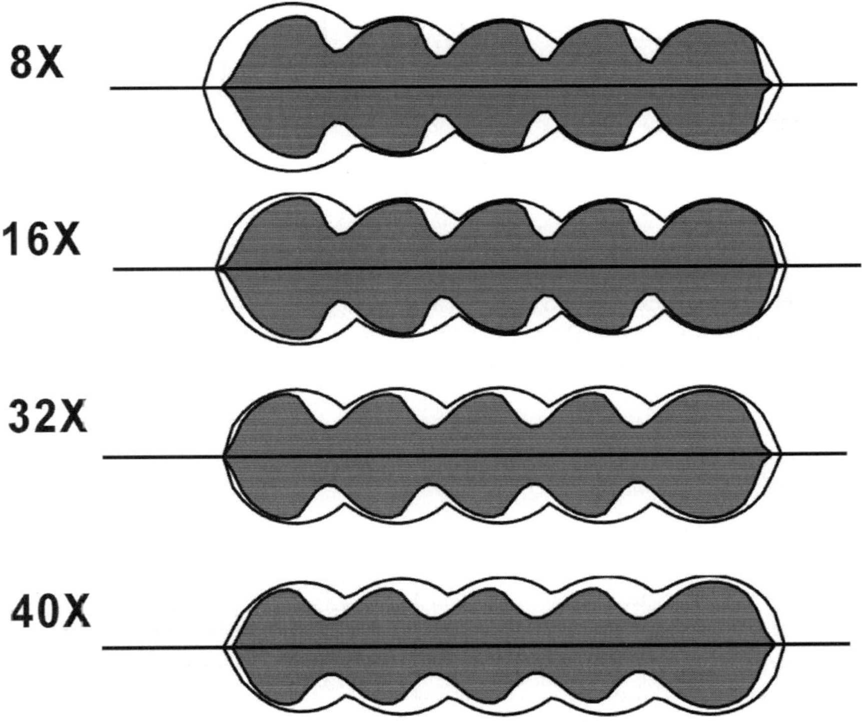

Figure 113. Mark-formation simulation results for a 10T mark written with the 2T write strategy at four different recording velocities (8×, 16×, 32× and 40× CD-RW). The solid line indicates the maximum molten area; the gray filled area is the amorphous mark after re-crystallization.

On the contrary, the temperature distribution is narrower and the achieved maximum temperature lower in case of a high recording velocity due to the involved shorter dwell time. A simple adaptation to the write strategy can compensate for this DC-heating effect. When the first write pulse is made shorter with a recording velocity-dependent factor, the mark shape can be improved.

Another recording velocity effect is seen in the re-crystallization in between two write pulses. Since the cooling gap between two write pulses decreases with increasing linear velocity, the time to cool down the stack is reduced. This leads to elevated stack temperatures when the next write pulse is fired. As a consequence, the re-crystallization during write, which is observed as a serration of the side edge, is more pronounced with increasing recording velocity. The resulting mark width is accordingly narrower.

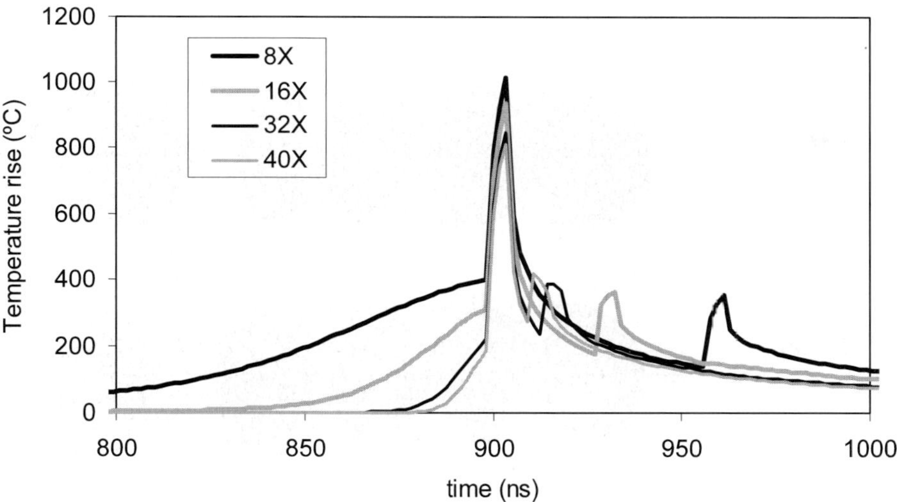

Figure 114. Temperature-time response to continuous laser power of 1 mW followed by two laser-pulses for different recording velocities. The temperature is plotted in the center of the phase-change layer.

When using a 2T write strategy, odd marks can be written with $N/2 \pm \frac{1}{2}$ pulses. Generally, it is chosen to use $N/2 - \frac{1}{2}$ pulses and extend the last pulse with Δ in order to melt an elongated part of the track. This can have consequences for the resulting mark shape. Computer simulations of an 11T mark are given in Figure 115 for three recording velocities (8x, 16x, 32x). The mark was written with 5 write pulses, the last write pulse being elongated with $\Delta1$ to obtain the required physical mark length. The length of the first write pulse was adapted to compensate for the DC-erase power. The serration of the sides of the marks becomes more pronounced for higher linear velocities, which is again due to the reduced time in between two write pulses.

The trailing edge of the odd mark suffers most from heat-induced re-crystallization. The elongation of the last pulse is required to write a physically longer mark, but the longer exposure time causes severe re-crystallization of the previously written amorphous area by direct heating. Also, more heat flows into the stack. This reduces the cooling down rate in the cooling period. Both effects cause severe re-crystallization of the mark tail during write. Back-growth becomes more severe at higher recording velocities because of the reduced cooling time between two adjacent write pulses.

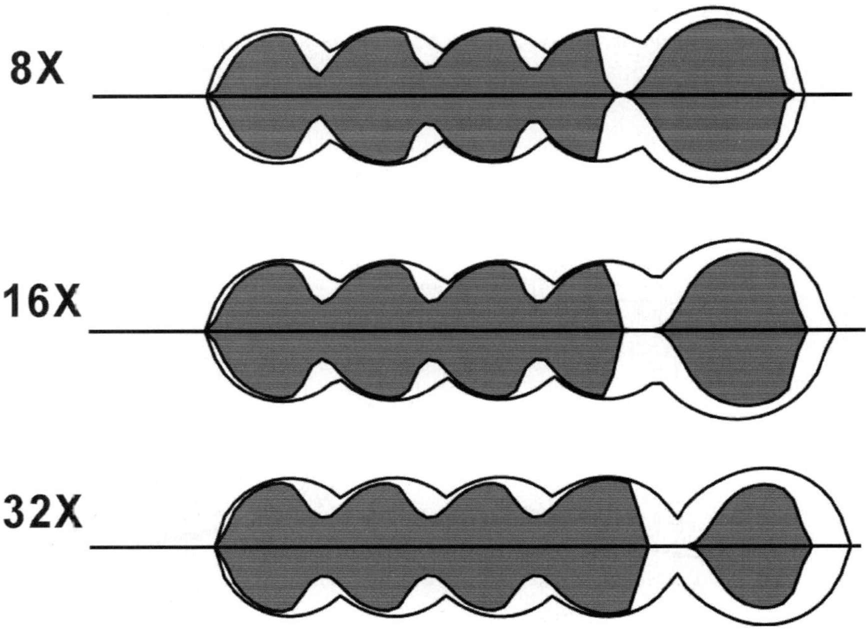

Figure 115. Mark-formation simulation results for an 11T mark written with the 2T write strategy at three different recording velocities (8×, 16× and 32×). The solid line indicates the maximum molten area; the gray filled area is the amorphous mark after re-crystallization.

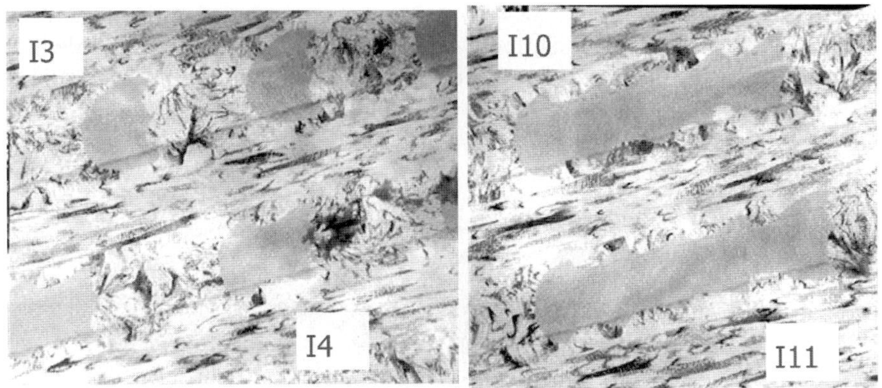

Figure 116. TEM pictures of marks written with the 2T strategy at 16× (DOW=10).

The numerically simulated effects are also recognized from Transmission Electron Microscopy (TEM) pictures of marks recorded with a 2T write strategy at 16x and 24x, see Figure 116 (16x) and Figure 117 (24x).

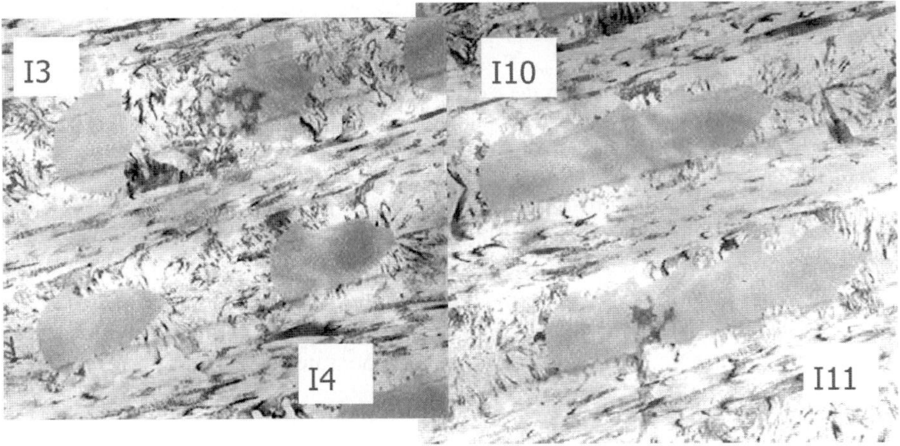

Figure 117. TEM pictures of marks written with the 2T strategy at 24× (DOW=10).

The main reason for the strange mark shape of odd marks is the elongated last write pulse and the concomitant heat accumulation in the stack. It is, therefore, worthwhile to look at the effect of the write pulse length in some more detail. Figure 118 illustrates the effect of the write pulse length T_{mp} on re-crystallization in the recording stack. Simulated marks shapes of a 10T mark are shown for $T_{mp}=1T$ and $T_{mp}=3/4T$.

The recording speed was 32x (38.4 m/s). The consequence of an elongated write pulse (T_{mp}=1T) is twofold. First, more heat leaks into the stack, which involves a reduced cooling rate and leads thus to more back-growth in the cooling down phase. Second, the cooling time in between two write pulses is reduced accordingly (since the 2T period is unchanged). Both effects lead to more re-crystallization and a very narrow amorphous mark.

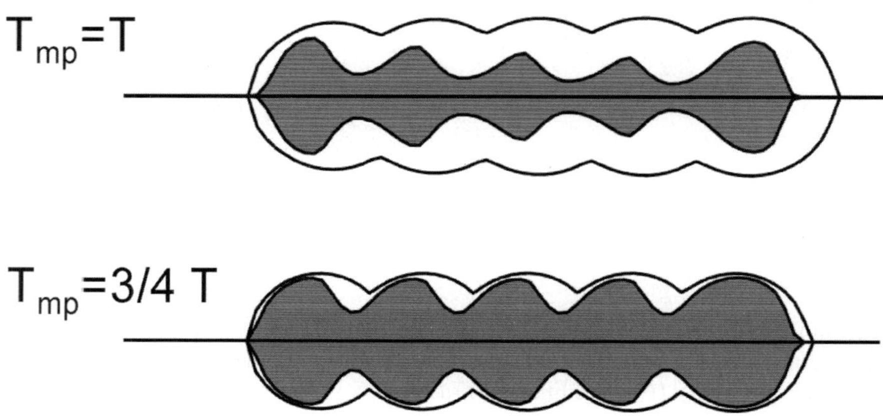

Figure 118. Mark-formation simulation results for a 10T mark written with the 2T write strategy: comparison between T_{mp}=T (upper plot) and T_{mp} =3/4 T (lower plot) at 32× recording. The solid line indicates the maximum molten area; the gray filled area is the amorphous mark after re-crystallization.

Extension of the last write pulse with $2\Delta1$ improves the mark formation somewhat, which is seen from the calculated melt-edges and mark-shapes (see Figure 119, only half of the mark shape is shown). The upper plot represents simulation results for an 11T mark written with a 2T strategy. The last write pulse is elongated with $2\Delta1$ in the results given in the lower plot. In addition, the length of the first write pulse was reduced with increasing recording velocity to compensate for the erase power effect.

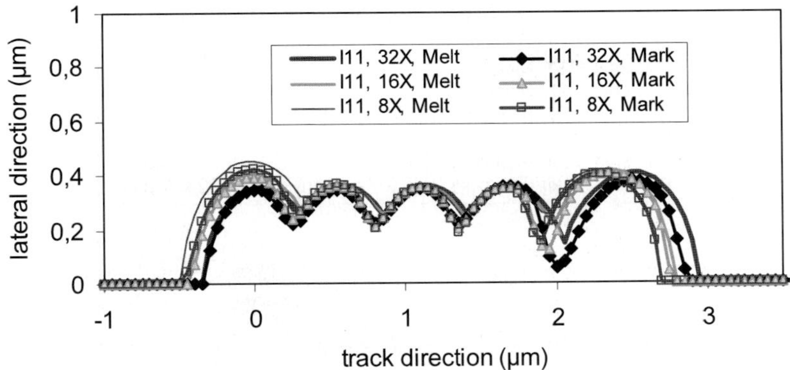

Figure 119. Calculated melt-edge (solid line) and mark edges (lines with symbols) of an 11T mark for three recording velocities written with the 2T strategy with one Δ1 elongation of the last write pulse (upper plot) and 2Δ1 elongation (lower plot, note that only half of the mark is plotted). Also the first write pulse is shortened with a recording velocity-dependent factor (lower plot).

One way to improve the tail of an odd mark is based on a modification of the write strategy with a staircase-shaped write pulse. [123] The temperature response of a staircase-shaped laser pulse is very steep compared to that of a block-shaped write pulse. Therefore, re-crystallization of previously written amorphous areas is noticeably suppressed while melting of the material still occurs due to the high-power levels in the end of the write pulse. The lower power stairs are still required to pre-heat the recording stack prior to melting. A simulation result for such a staircase strategy is given in Figure 120. It can be seen that the write performance is somewhat improved. If a relatively short additional write pulse is applied next to the staircase pulse, such that the last 3T-part of the 11T mark is written with two write pulses, on of them being the staircase, the back-growth is further suppressed. Results for such a write strategy are also given in Figure 120. The constriction in the tail of the mark, as appears for the other cases, is almost completely suppressed.

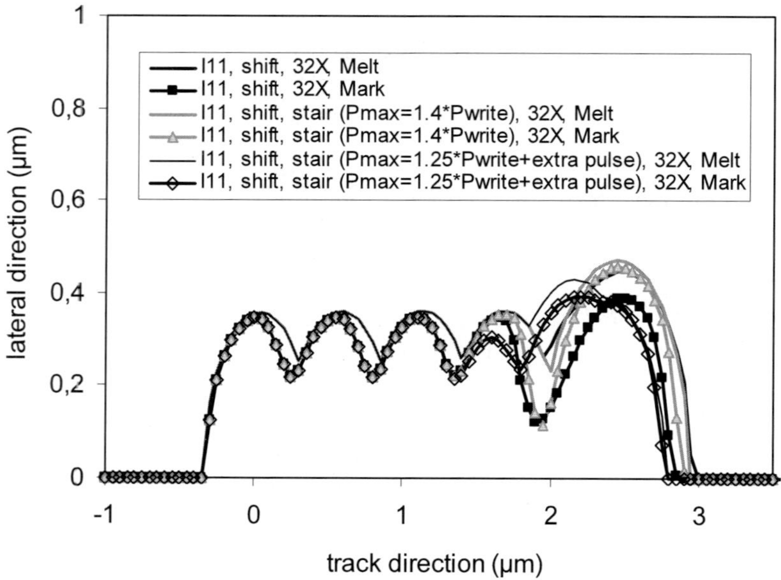

Figure 120. Calculated melt-edge (solid line) and mark edges (lines with symbols) of an 11T mark at 32×recording speed for two improvements of the write strategy for the tail of the mark: stair case and stair case with additional short pulse.

Mark formation @ 1T write strategy

The 1T (or N-1) write strategy with short write pulses was proposed as the default write strategy for 16x CD-RW recording. The effect of duty cycle is illustrated in Figure 121. The clock time T was similar for both simulations, the pulse width was $T_{mp}=1/2T$ ($P_w=26$ mW, $P_e=8$ mW, 50% duty cycle) and $T_{mp}=1/4T$ ($P_w=38$ mW, $P_e=8$ mW, 25% duty cycle). The longer write pulse causes severe re-crystallization, similar to the case of the 2T write strategy with long write pulses. The 1T strategy gives only good results if short write pulses are used.

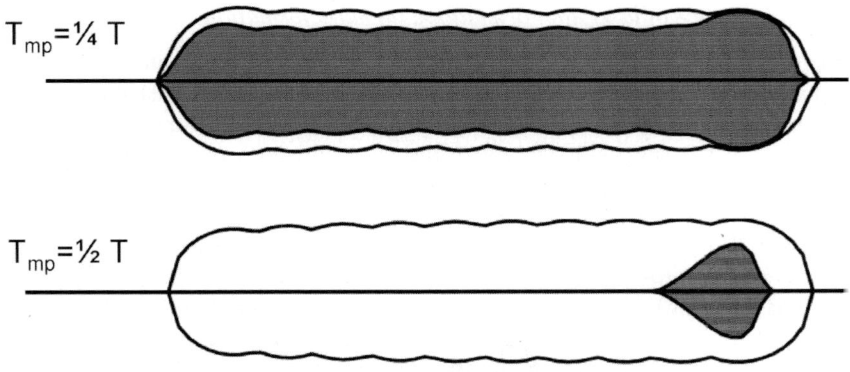

Figure 121. Mark-formation simulation results for a 10T mark written with the 1T write strategy at 32× recording: comparison between $T_{mp}=1/4T$ (25% duty cycle, upper plot) and $T_{mp}=1/2\,T$ (50% duty cycle, lower plot).

TEM pictures of marks written with a 1T write strategy compare again well to the computer simulations, see Figure 122.

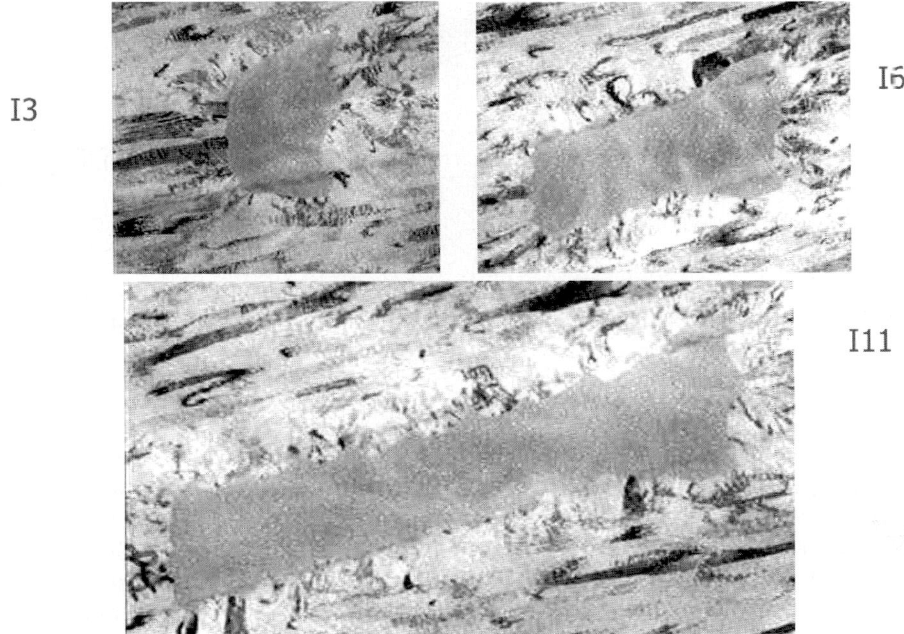

Figure 122. TEM pictures of marks (3T, 6T and 11T length) written with the 1T strategy at 16x.

5.2.6.   *Influence of laser spot orientation on mark formation*

A beam shaper is used in DVD and BD optical paths to compensate for the ellipsoidal shape of the laser beam coming out of the laser diode. In CD-RW optics, the beam shaper is omitted, the resulting spot has indeed an ellipsoidal shape. The orientation of the spot with respect to the track direction (grooves) then becomes important, especially in the case of high-speed recording.

In practice, three different spot orientations can be distinguished. The case with the long axis of the ellipsoid oriented along the track is denoted as TOS (oriented in the tangential direction). The case of perpendicular orientation to the track direction is called ROS (oriented in the radial direction). An orientation in between, with the axis at an angle of 45 degrees with respect to the track direction is denoted as DOS (oriented in the diagonal direction). Schematics of the DOS and ROS orientations are given in Figure 123.

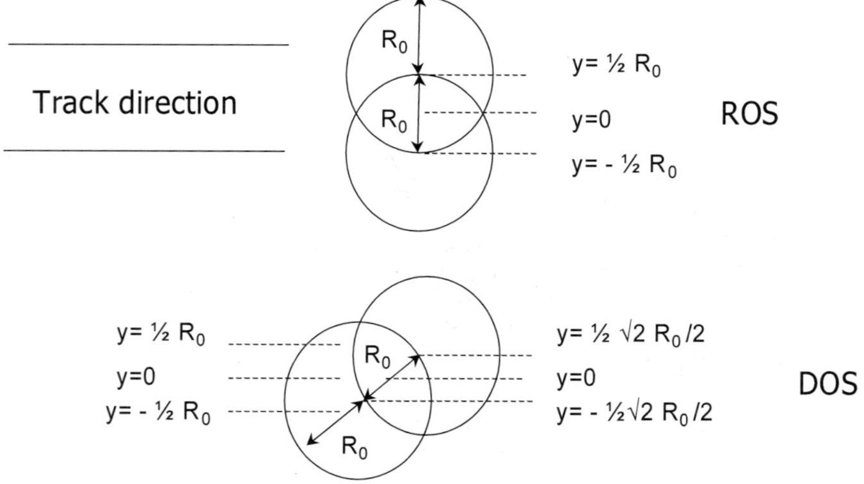

Figure 123. Schematics of a DOS and ROS orientation of the laser spot in case no beam shaper is used to compensate for the ellipsoidal shape of a laser beam coming out of a laser diode.

One of the issues foreseen in high-speed phase-change recording is the drive-to-drive playability of discs. If, for example, a disc is recorded with a ROS-recorder, the written marks will be broad. If the disc is subsequently overwritten with a DOS recorder, which has a narrower laser spot along the data track direction, incomplete erasure of the amorphous marks may occur, in particular in case of high recording speeds. Furthermore, the erasability of a system with DOS orientation may be different from that with a ROS orientation. This can be understood from a dwell-time point of view. The projected distance along the track of a DOS spot is larger than that of a ROS spot. Experiments indeed indicate that a DOS spot is beneficial for erasing old data.

Computer simulations have been performed for a ROS and DOS spot to verify the improved erasability of the DOS orientation. The ellipsoidal laser spot was created from superposition of two Gaussian shaped circular spots at R_0 distance; see the sketch in Figure 123. The resulting spot shape is illustrated in Figure 124. This superposition results in broadening of the normalized spot along the axis of superposition, while the other direction is unaffected. Also, the superposition at R_0 distance ensures a flat area in between the centers of the circular spots. It should be noted that the superposition at R_0 distance gives a somewhat broader laser spot than typically encountered in real CD-RW applications (deviations of about 10% in axis length are quite common). However, for the sake of a qualitative interpretation this is not a problem.

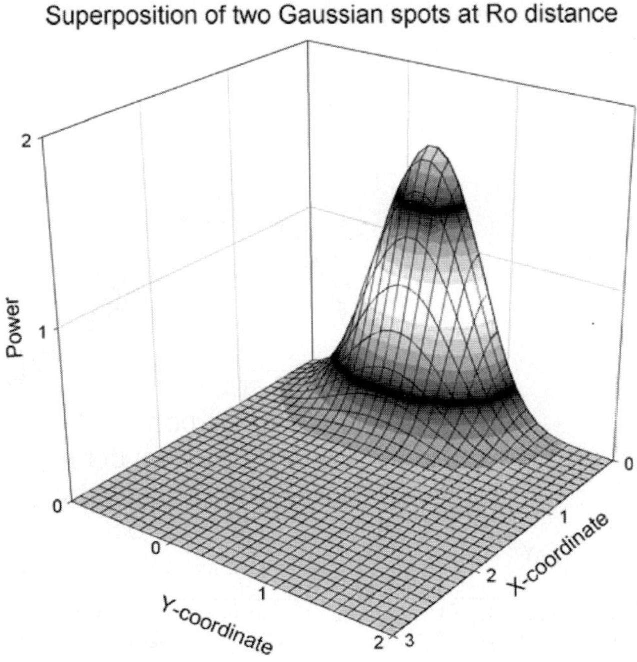

Figure 124. Spot size that results from superposition of two Gaussian laser spots at R_0 distance. The y-axis is the axis of superposition.

The temperature response to a continuous laser power is given in Figure 125 for a DOS and ROS orientation at 16x recording speed. Temperature profiles in the center of the phase-change layer are plotted at three lateral locations in the disc, namely $y=-\frac{1}{2}R_0$, $y=0$ and $y=\frac{1}{2}R_0$ (see the sketch in Figure 123 for the lateral positions). The ROS spot has a symmetric shape. Therefore, the temperature profiles at $y=-\frac{1}{2}R_0$ and $y=\frac{1}{2}R_0$ are similar. The spot shapes result in different temperature responses, each having a different delay time. To allow for a comparison of the differences in dwell time, which is actually the time available for erasure of old amorphous marks, the temperature profiles have been slightly shifted in time such that the heating-up flanks coincide with each other. As can be seen, for the DOS orientation the dwell time at $y=0$ is larger than at $y=\frac{1}{2}R_0$. In addition to the higher maximum temperature, erasure will start in the center of the track. The temperature profiles calculated for the DOS case are asymmetric. The temperature profile at $y=0$ has the longest dwell time and highest maximum. Furthermore, the optical spot will reach first location $y=\frac{1}{2}R_0$ and will preheat the recording stack. Therefore, location $y=-\frac{1}{2}R_0$ will experience a preheat effect, resulting in a longer effective dwell time (this can also be deduced from the sketch in Figure 123). Similar observations are made for a higher recording speed of 32x, results are shown in Figure 126. The higher recording speed results in shorter dwell times and, hence, lower maximum temperatures.

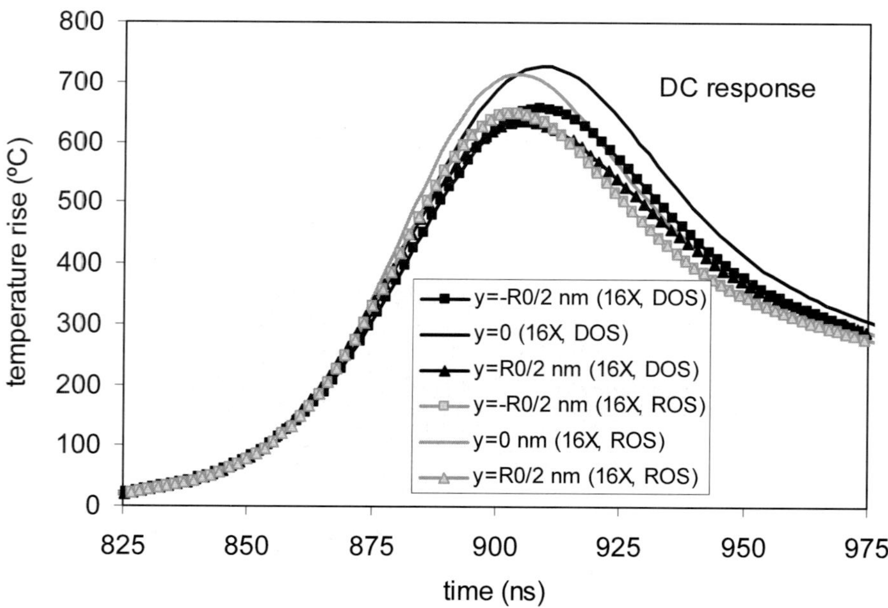

Figure 125.Temperature-time responses to a continuous laser power for a ROS and DOS orientation at 16× recording speed.

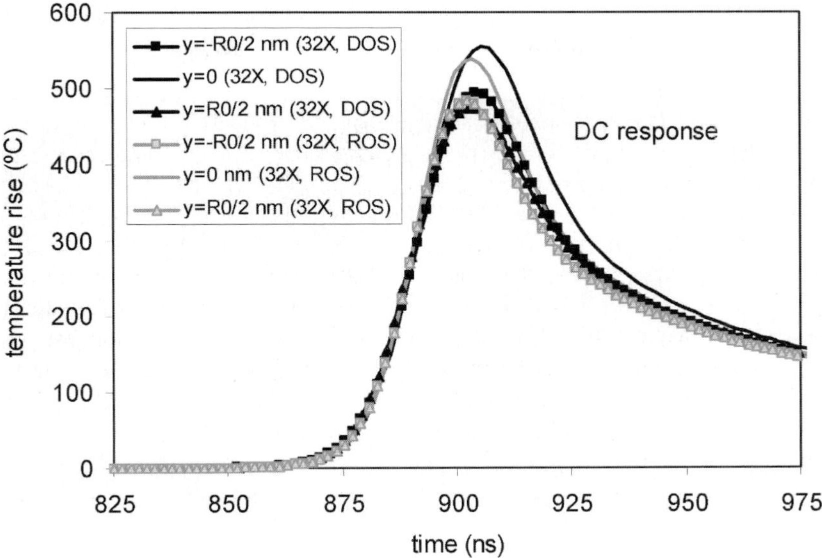

Figure 126. Temperature-time responses to a continuous laser power for a ROS and DOS orientation at 32× recording speed.

Calculated mark shapes for a ROS and DOS orientation of the laser spot are shown in Figure 127 for 32x recording. As expected, the serration of the side edge is asymmetric for the DOS spot.

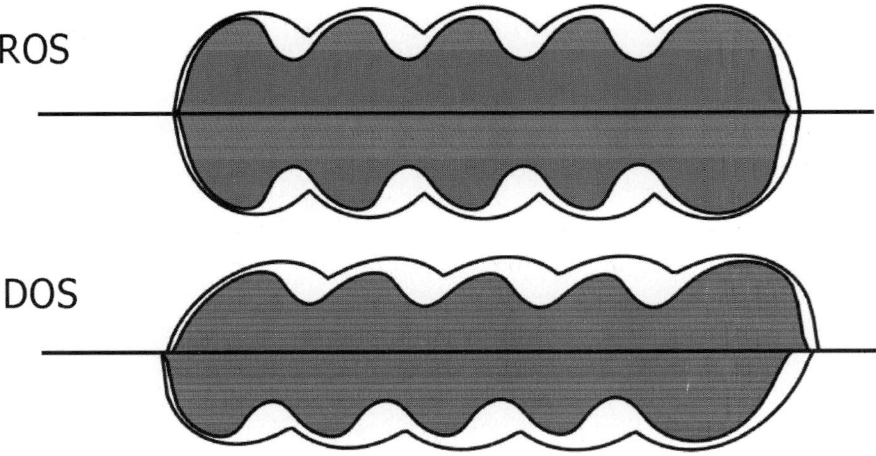

Figure 127. Mark shapes of a DOS and ROS orientation of the laser spot at 32× recording.

TEM images of marks written with a DOS spot and a ROS spot at 16x recording speed (2T write strategy) are given in Figure 128 and Figure 129, respectively. The re-crystallization in the trailing edge and at the sides of the marks is clearly visible.

Figure 128. TEM pictures of marks written in a CD-RW disc written with a 2T write strategy at 16x (DOW=0, DOS-spot).

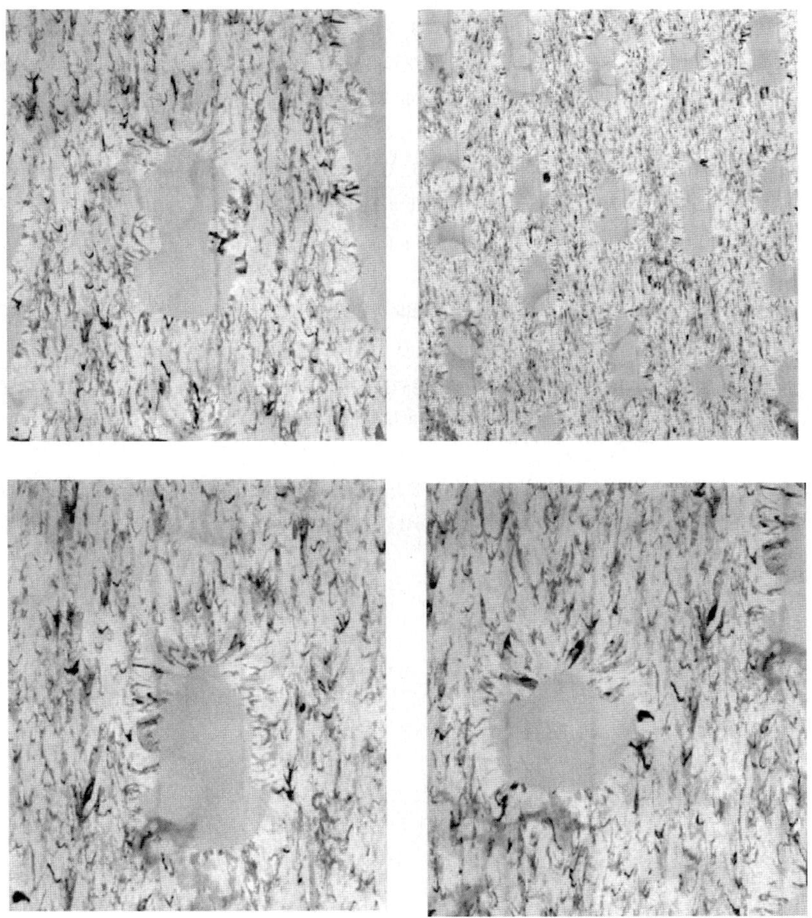

Figure 129. TEM pictures of marks written in a CD-RW disc at 24 X recording (DOW=0, extreme ROS-spot).

5.3.   References of chapter 5

[117]   K. Yokoi, I. Aoki, US-patent US5732062, 24 March 1998.

[118]   N. Nobukuni, M. Horie, European patent EP1182649(A1), 27 February 2002

[119]   H.J. Borg, M. van Schijndel, J.C.N. Rijpers, M. Lankhorst, G. Zhou, M.J. Dekker, I.P.D. Ubbens, and M. Kuijper, 2001, Phase-change media for high numerical aperture and blue wavelength recording, Jpn. J. Appl. Phys., Vol. 40, pp. 1592.

[120]   E.R. Meinders, H.J. Borg, M.H.R. Lankhorst, J. Hellmig, and A.V. Mijiritskii, 2002, Numerical simulation of mark formation in dual-stack phase-change Recording, J. Appl. Phys., Vol 91, No 12, pp. 9794-9802.

[121]   A.V. Mijiritskii, J. Hellmig, H.J. Borg and E.R. Meinders, 2001, Development of recording stacks for a rewritable dual-layer optical disc, Jpn. J. Appl. Phys., Vol 41, 1668.

[122]   L. van Pieterson, J. C. N. Rijpers and J. Hellmig, Jpn. J. Appl. Phys. **43** 4974 (2004)

[123]   E.R. Meinders and J. Hellmig, Re-crystallisation-controlled write strategy for recording in dual-layer phase-change stacks, PHNL020780, 2002, WO2004017308 A1, 2004

[124]   K. Narumi, S. Furukawa, T. Nishihara, H. Kitaura, R. Kojima, K. Nishiuchi, N. Yamada, ISOM2001 Technical Digest, 202.

Philips Research Book Series

1. H.J. Bergveld, W.S. Kruijt and P.H.L. Notten: *Battery Management Systems.*
    2002   ISBN 1-4020-0832-5
2. W. Verhaegh, E. Aarts and J. Korst (eds.): *Algorithms in Ambient Intelligence.*
    2004   ISBN 1-4020-757-X
3. P. van der Stok (ed.): *Dynamic and Robust Streaming in and between Connected
    Consumer-Electronic Devices.* 2005   ISBN 1-4020-3453-9
4. E.R. Meinders, A.V. Mijiritskii, L. van Pieterson and M. Wuttig: *Optical Data
    Storage.* Phase-Change Media and Recording. 2006   ISBN 1-4020-4216-7
5. S. Mukherjee, E. Aarts, R. Roovers, F. Widdershoven and M. Ouwerkerk (eds.):
    AmIware. Hardware Technology Drivers of Ambient Intelligence. 2006
                                                  ISBN 1-4020-4197-7